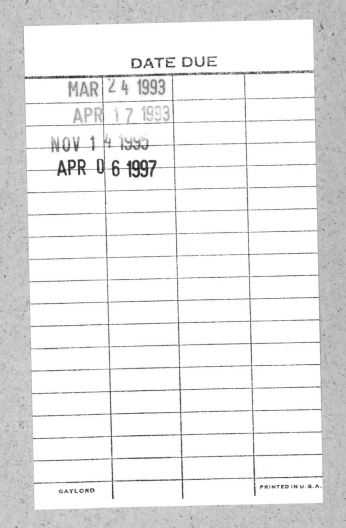

The Book of Where
or How to Be Naturally Geographic

The Book of Where
or How to Be Naturally Geographic

by **NEILL BELL**

illustrated by **RICHARD WILSON**

Little, Brown and Company

Boston   Toronto   London

This Brown Paper School book was edited and prepared
for publication at the Yolla Bolly Press, Covelo, California,
during the spring of 1981. The series is under the supervision
of James and Carolyn Robertson. Editorial and production
staff: Dan Hibshman, Barbara Youngblood,
and Joyca Cunnan.

Library of Congress Cataloging in Publication Data

Bell, Neill.
The book of where, or How to be naturally geographic.

(A Brown paper school book)
Summary: Text, illustrations, and suggested
activities outline such basic concepts of geography
as scale, maps, the globe, continental plates, and oceans.
1. Geography—Juvenile literature.
[1. Geography] I. Wilson, Richard, 1936-
II. Title.
G133.B43   910   81-19315
ISBN 0-316-08830-7   AACR2
ISBN 0-316-08831-5 (pbk.)

First edition. Published simultaneously in Canada
by Little, Brown & Company (Canada) Limited.
Printed in the United States of America.

Library of Congress Cataloging in Publication Data

HC: 10 9 8 7 6 5 4 3 2   BP
PB: 10 9 8 7 6 5 4   BB

This book is dedicated
to all those brave adventurers who step
from the familiar into the unknown—
from the edge of a continent
or from a front porch.

What's in this book

The Book of Where
or How to Be Naturally Geographic

How to Be Naturally What?

It's tough not knowing your way around in the world. For example, you could be listening to your radio and hear, "A strong tremor shook the Philippines today," or "This week a determined group of young climbers scaled one of the highest peaks in the Andes," or "The world's largest naval vessel sailed through the Golden Gate a few hours ago."

If you know very little about the world and its places, you wouldn't have a very good idea of what has happened, where it happened, and why it is important to you. You might think that the Philippines are a new rock group or the people who live in a large east coast city or maybe even an all-you-can-eat vegetarian restaurant. You wouldn't know that the Philippines is a group of more than 7,000 tropical islands lying between the Pacific Ocean and the South China Sea.

You might guess that the Andes are giant fish or a bar located on the top floor of a skyscraper—instead of a towering mountain range in South America. The Golden Gate may seem more like the heavy door that keeps drunken sailors locked in jail rather than the entrance to San Francisco Bay.

Without knowing something about where in the world places like these are, you won't be able to make much sense out of the news. More and more, as we are all finding out, events that happen thousands of miles away have an important effect on our lives.

Closer to home it's even more important to know your way around. If the radio message had been about a record store that will be giving away free samples this weekend at their new location at Third and Market, you may be very interested. But if you don't know where Third and Market is or how to find it, you and your stereo are just out of luck.

Another reason to know where in the world you are is so that you don't get lost. You don't have to be in the wilderness to lose track of where you are or where you are going; and no

7

matter where you are, getting lost can be very frustrating, and even a little scary.

Okay, so it's a big world and a complicated one. A lot of grownups who have been around a long time don't know Honshu from Timbuktu any more than you do. But you can make the world seem a more interesting place and add some excitement to your life if you just take the time to know it better.

You may even surprise yourself with how much you already know—after all, you are a real expert in your own part of the world. To find your way in this book you won't need much more than a pencil (with a good eraser), some paper, a few maps, and things you can easily find around your house. Keep reading and you'll get there in no time.

Geography:
the study of place
that hopes to answer the
question of where.

CHAPTER ONE
Starting at Home

There is one part of the world you may know better than anyone else—your own home. Chances are that you can find your way blindfolded to the refrigerator from anywhere in the house.

Just how well do you think you know your house? Imagine that you are standing in the middle of your living room and you want to get the gum you hid under your bed from your sister. Could you get there with your eyes shut?

Let's see if you are bluffing. First make a list of ten or so places in your house, such as the kitchen sink, the bathtub, a clothes closet, the place where you keep the TV set, the dining room table, the front door. When you have listed the places, have a friend tie a blindfold over your eyes (or you can just close them if that feels better). Then go from one place to the other, working down the list until you have been to all of them.

Hints: You may want to move things out of the way before you begin this game of Blind Kid's Bluff so that you don't destroy them or yourself. Roller skates, balls, and banana peels are especially bad things to step on when your eyes are closed. Walk slowly. And if you lose your way, open your eyes just long enough to get your bearings again. Or instead of opening your eyes, just stand still and listen. Sometimes sounds will help you get your bearings. Good luck.

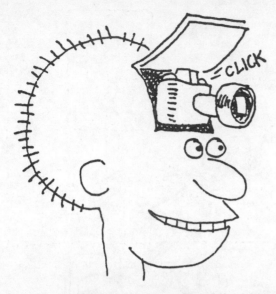

Camera in Your Head

Just how well you are able to find your way around with your eyes closed depends on how well you take pictures with the camera in your head. That's right, you may not hear the shutter working, but you are storing pictures up there that you can use later on to find your way around places you have been before. People who are very good at doing this—who can do it after seeing a place only once—are said to have photographic memories.

Actually the camera in your head doesn't have to catch every detail to work well for most purposes. It should be sharp enough to let you see and remember important reference points, such as doorways and large pieces of furniture, and to help you know which direction to turn and about how far you have to go before changing directions. If your camera doesn't do this very well, your nose could temporarily become a part of the wall. (Ouch!)

Even the best cameras don't take good pictures without the proper exposure, so if you didn't do so well the first time, look around carefully and take another snapshot with your noggin.

Then try finding your way around again. With a little practice you could become a blindfolded wonder. You could even get this trick to pay off for you. Bet your mom or dad that you can find your way to the cookie jar from your bedroom with a blindfold on—if you get a prize. Cookies and ice cream may be suitable prizes. Then use the pictures in your mind to find your way to the feast.

10

There are other ways to test how well your photographic memory works and sharpen up those pictures at the same time.

You often see these shapes with words on them. Can you remember which words are usually on them? They are traffic signs that you have seen many times.

Answers: 1. Stop, 2. Yield, 3. Parking, 4. Speed Limit.

Here are a few more things to try to recall, things mostly found around your home. Don't look without taking a guess first.

1. Is the hot water faucet on the left or right?

2. Which hand of the clock moves faster—the big one or the little one?

3. Do you twist a jar lid or bottle cap clockwise or counterclockwise to open it?

4. Does a record turn clockwise or counterclockwise on the stereo?

5. Are your shirt buttons sewn on the left side or the right side?

6. Is the refrigerator door handle on the left side or the right side?

Answers: 1. Hot is usually on the left; 2. The big (minute) hand moves faster; 3. Counterclockwise; 4. Clockwise; 5. That depends on whether the shirt was made for girls (buttons on the left side) or for boys (buttons on the right)—who knows why!

6. Consult your own refrigerator, since they are made both ways.

You might try to test your parents' or friends' mental cameras with the questions you just answered or with your own questions.

One last toughie for you. Can you remember what your mom or dad is wearing right now? Guess, then check your answer.

Flies' Eyes

Now that you are an expert in knowing what your house is like—even with your eyes shut—let's take a look at what it would look like to someone else. Like a bat, or maybe a fly. Your kitchen looks very different to a fly on the ceiling than it does to you.

Actually, the fly's-eye view is more of a map than a picture, but it gives you some idea of what the room would look like from above. That's how maps often look, from a bird's-eye or a fly's-eye view.

Maps not only help people find their way around, they also help the map-maker get a clearer picture of the place being mapped.

Your own house is probably small enough so that people wouldn't need a map to figure out how to get around in it. (Some castles and large mansions actually *are* big enough to get lost in, if you don't have a map to help.)

You may notice that some things have been left out of the view from the ceiling. Like most maps, it has been simplified; only the things that don't move have been left in. That's why the cat isn't there.

Mapping your house is a good way to see just how much you really know about it.

Try making a map of your home that shows only the rooms inside it. Unless it is round, or some other nonrectangular shape, it should look like a bunch of boxes stuck together, with each box representing a room.

After you have drawn your map, take a good look around your house. Do the boxes fit together in the right places? Are they all the same size, or do they appear larger or smaller according to the sizes of the rooms they represent? This is when you may want to use the mapmaker's friend—the eraser.

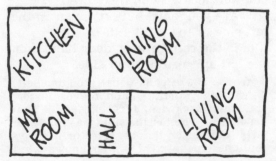

What Scale?

If you drew all the rooms in about the right proportion relative to one another, then you drew them to the same scale. The what? Scale. The drawings below give you an idea of scale. An ant should be a lot smaller than an anteater, right? So one of the drawings below is drawn to scale, and the other one is not.

Maps are the same way. If things are kept on the same scale, a person using the map will have a better idea of the way things really look. Like the anteater's tongue, things work better when they are the right size.

Now let's see how good you are at scaling. The objects on the right are drawn to three different scales—A, B, and C. See if you can arrange them in three groups according to scale.

Answers: A. 1, 5, 7; B. 2, 6, 8; C. 3, 4, 9.

The idea of scale is important in mapmaking and reading because distances represented on paper should be sized according to the way they are in real life.

To make the map of your room accurate, you will need to measure your room. Using a ruler or yardstick, measure along each wall and record your measurements.

Let's use a scale of 1 inch = 2 feet. This means that every inch of line on your map represents 2 feet in your room. (A wall 12 feet long is shown by a 6-inch line, for example.)

Using the edge of a book or card to keep corners square, measure and draw each wall on paper. Is your map a good fly's-eye view of the room?

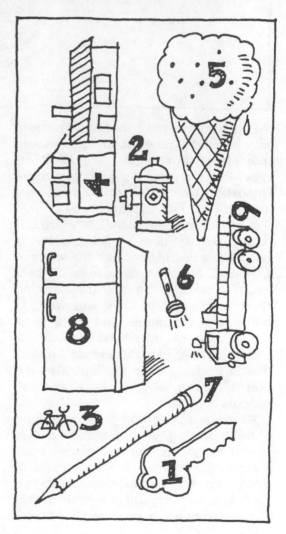

CHAPTER TWO

The Streets Where You Live

Someday you may need to know your way around Timbuktu, an ancient and interesting-sounding place in West Africa. But for now, it's probably more important that you can find your way around your own neighborhood.

What is a neighborhood? It's difficult to say exactly, because neighborhoods don't have boundaries like the walls of a house. Most people consider their neighborhood to be the area around their home, places they can walk to within a few minutes. But there is no set size. A neighborhood usually includes not only the houses and apartments where people live but also the stores, parks, shops, churches, and schools.

Perhaps a good way for you to think about how big your neighborhood is would be to include the places near your home where you spend most of your time. The general area where you go to school, play, and feel comfortable just hanging out is your neighborhood.

Let's see just how much you really do know about the streets where you live.

Places You Go — Places You Know

Every neighborhood has some places where kids spend a lot of time playing—where parents and other grownups don't ordinarily go. If you want to meet friends after school, maybe you just say, "See you at the big walnut tree," or "Let's meet by the old water tank." Your friends know where you mean, but your parents might not.

14

An alley cat probably knows a few things about your garbage cans that *you* would never guess, or even care about. If the cat could talk, it would tell you the neighbors finally threw out leftover meatloaf and that your kid brother didn't eat all of his fish last night.

Even so, there are some places that practically everyone who lives in your neighborhood knows. These could be big stores, public buildings, parks, or intersections of main streets. Neighborhood reference points are the landmarks that people use to help find their way around.

Here's one way you can locate the reference points in your neighborhood. Find a corner where a lot of people pass by on foot. Ask for directions to some place you know is on the other side of the neighborhood. Take along a pencil and a piece of paper and write down the places these people mention as they give you directions. If several people mention the same reference points, make marks to add up later to find out the number of times those places were named.

You will need to ask at least ten cooperative passersby to get a good sample. A few more would be even better. If people catch on to the fact that you aren't really desperate for the directions, just tell them that you are doing a survey.

The survey should give you the following information: (1) which places people most often use as reference points, (2) whether different people give directions in the same way (you can try to figure out why not if they don't), (3) how many people know how to get from here to there, and (4) how many people are bugged by kids who ask for directions.

Short Circuit

Now let's see how sharp you are at remembering just where things are close to home. On the following page are illustrations of things you are likely to find in your neighborhood. All you have to do is plan a trip that will take you by each one. The trick is to list them all in order so that you make the shortest possible trip, without having to backtrack. What should you head for first? Can you make the shortest circuit without a mistake?

If there's more than one of some of these things in your neighborhood— two gas stations, for example—pick whichever one keeps the trip shorter.

The only way for you to see how well you did with your list is to go check it for yourself. Did you go to any place out of order, or did you forget ones that are closer to home?

Once you have made any necessary corrections on your trip list, try this short-circuit experiment on your parents or a friend. It's pretty safe to bet they won't do as well as you did.

15

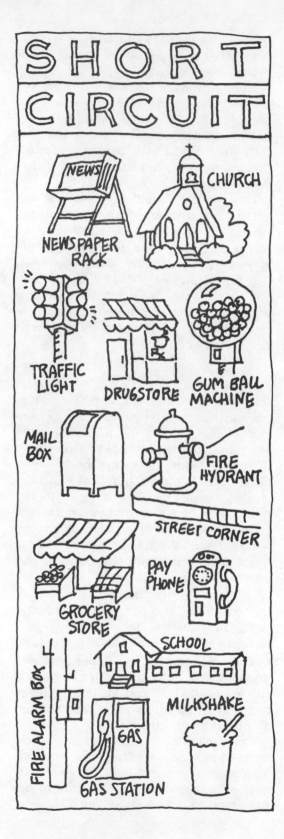

The Name Game

Names— you find them everywhere. If you didn't, it would be hard to find your way around. That would be especially true in today's big cities, with their hundreds of streets and thousands of buildings.

Even neighborhoods have names. Some are supposed to sound like nice, peaceful places to live; there are Pleasant Views, Sunnysides, and Sunset Hills. But there are also older names that sound more interesting and funnier—Skunk Hollow, Tight Squeeze, Turkeytown, and Cheapside.

The names of streets are very important in finding your way around. See if you can list the names of all the streets within five blocks of your house. If that is too easy for an old pro like you, try listing all the streets within ten blocks. Or if your school is a long way away, list the names of all the streets you have to cross to get there.

You may have noticed that not all streets are called "Street." There are other names they go by, such as "Avenue" or "Road." Below are some abbreviated—shortened—versions of names that are often given to streets. You are likely to run into them on signs and maps where space is limited. Can you figure out what they stand for? Pl., Blvd., Wy., Ct., Dr., St., Ln., Ave., Cir., Ter., Hwy., Rd., Sq., and Pkwy.

MOST TOWNS AND CITIES HAVE STREET PATTERNS THAT MAKE THEM LOOK LIKE GIANT WAFFLES.

The raised parts are the streets, and the holes (where the butter melts into the syrup) are the blocks of houses, other buildings, and parking lots.

The streets in this wafflelike grid pattern are arranged in two groups. Those belonging to one group are parallel—they go in the same direction and don't cross each other. They do cross all the streets in the other group, just like the ridges on a waffle.

Fortunately for us, the people who named the streets often matched this grid pattern with their naming pattern. This makes it easier to remember the street names—if you know the pattern.

Answers: Place, Boulevard, Way, Court, Drive, Street, Lane, Avenue, Circle, Terrace, Highway, Road, Square, Parkway.

Here are some examples of street names that have patterns. Look for the pattern in each set, and then check the answers that follow.

Streets	Cross Streets
Maple St.	23rd St.
Laurel St.	24th St.
Elm St.	26th St.
Live Oak St.	27th St.

The first group of streets is named after trees, while the cross streets are numbered in order. What happened to 25th St.? It might have been cut off by a park or gotten lost in some other part of town.

Franklin St.	Vermont Ave.
Gallatin St.	Arizona Ave.
Hamilton St.	Michigan Ave.
Ingham St.	Rhode Island Ave.

The first group of streets is in alphabetical order. Those streets also carry the names of government leaders of nearly 200 years ago (but who can remember that far back?). The cross streets all have state names and are called avenues.

Bluebird Blvd.   Meadowlark Ter.
Sparrow Ln.   Hummingbird Dr.
Woodthrush Cir.   Cardinal Ct.
Robin Rd.   Woodpecker Pkwy.

One thing you can see here is that the people who named these streets were bats about birds. Some neighborhoods have this kind of naming pattern, where all streets running in both directions have similar names. There is one other pattern that is very hard to spot—look again closely and say the names out loud. Do you hear it? The streets in one group have two syllables (Rob-in, Spar-row), while the cross streets have three (Mead-ow-lark, Car-di-nal). Pretty sneaky, huh?

Does your neighborhood have patterns like these? If you have numbered or lettered streets, the pattern is easy to spot, and you can usually guess the name of the street that comes next.

Numbers take the name game one step further. When used with a street name, they tell you exactly where a building can be found. It's a lot easier to get a letter delivered when you address it "248 Marfunkel Place" than when you address it "The pink house on Marfunkel Place, three doors down from the brown church."

Once you figure out how the numbering system in your neighborhood works, chances are you'll know the pattern for the rest of the town.

NUMERAL OHNO, DETECTIVE

A STORY YOU HELP WRITE BY DOING IT

I began this investigation on the corner closest to my house. I hung around on the corner for a while just to case the block. Since the coast was clear, I started walking down the block toward my house. Out of the corner of my eye, I caught a glimpse of the first house number. Although it was getting dark, my super-vision let me see that the number was _____.

As I continued strolling casually down my street, I noticed that the numbers were getting (larger, smaller —you pick the correct word here). And although I was really playing it cool, I still caught the number of the last house on the block, _____.

When I got to the corner, I saw that a big ten-ton garbage truck with the meanest-looking driver this side of Cleveland was waiting for me to cross the street. I knew that if I put so much as one toe into that intersection, it would be mashed flatter than Tuesday by the big skins on that overgrown trash can's wheels. So I just turned the corner, whistling to myself and saw that the first house on the cross street had the number _____.

I kept walking down the side street, watching the numbers getting (larger, smaller) as I kept the eyes in the back of my head fixed on the trash masher. The last house number I saw before I reached the next corner was _____.

I knew I wasn't going to cross the street there either. On the opposite corner sat a guy in a motorized wheelchair. To anyone but a trained detective, he looked like a sweet, little old lady in a granny bonnet, but my piercing glance picked up the fact that she had forgotten to shave and was wearing combat boots. So I just turned the corner again and headed down the street that runs in back of my house, noting that the first number on the block was _____.

The rest of that block whizzed past, the numbers getting (smaller, larger) as I walked swiftly toward the corner. I was barely able to glimpse the last number on the block, _____, when I saw something that chilled my blood.

There on the other side of the street, in plain view, was the fiercest-looking pack of killer worms I had ever seen. I could even hear 'em gnashing their teeth as I turned the corner and headed for the place I had begun this investigation. I pulled the collar on my trench-coat higher as I snuck a look at the first house number on this street, _____.

The numbers got (smaller, larger) as I began to sweat, feeling the burning eyes of the killer worms heating the back of my neck.

But even though my life wasn't worth a plug nickel just then and I could sense danger all around, that didn't prevent me from taking in one final fact — the number of the last house before the corner, which was _____.

I went around the last corner and headed for home with my notes. Although they tried every trick to stop me from finding out their numbering system, I've gotten everything I need to figure it all out for myself. I bolted through the door like an elephant heading for a peanut vendor. I wasn't really scared, but I'm glad you came along, kid. Now all we gotta do is crack this code. Come to think of it, why don't you try to do it yourself.

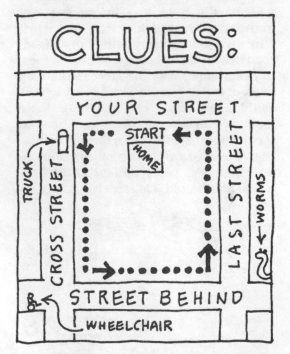

1. If you know which way the numbers run, higher or lower, in one direction on a street, you know the way they will run on all the blocks of that street.

2. Every cross street usually begins a new series of 100 numbers — if you are on the 700 block of a street, the next block will be either the 800 or 600, depending on how the numbers run.

3. Parallel streets (those that run in the same direction and don't cross each other) usually have the same 100 block numbers when they reach the same cross streets.

4. Take a look at the numbers you wrote down — are they odd (ending with a 1, 3, 5, 7 or 9) or even (ending with a 2, 4, 6, 8, or 0)? Does this have anything to do with which side of the street they are on? Think about it.

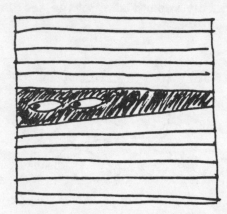

Okay, kid. I just looked out through the blinds — the garbage truck is on its way to the dump, the little old lady is at the drag strip, and the killer worms are trying to eat their way to China. Make a couple of guesses about the numbers you would find if you crossed the street and walked down a couple of blocks. Do you see the pattern? Go ahead, take a chance. Pretty soon you'll get the big picture. This town is a knockover. Gotta go — my mom is calling me. See ya, kid.

THE END

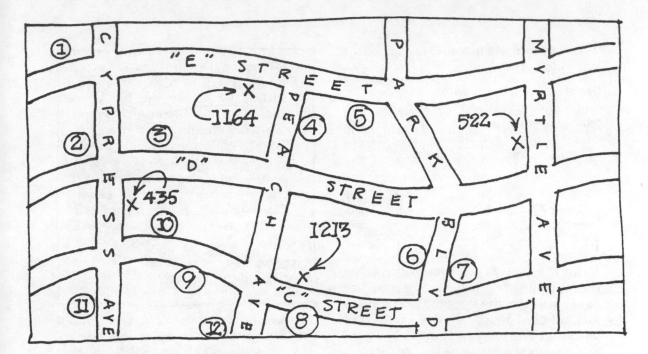

Are You Just As Good in Another Neighborhood?

Now you have a chance to prove that you can do just as well in a neighborhood you don't even know. The only things you have to work with are the street names and four numbered addresses. From these you have to match six addresses given below with the correct circled numbers on the map. Can you do it? Here goes.

423 Park Blvd.   1075 E St.
1156 C St.   527 Peach Ave.
340 Cypress Ave.   1117 D St.

If you got all of these right, you won't have much trouble finding street addresses in your town, or in other towns for that matter. If you missed most of them, you may want to practice before applying for a job as a mail carrier.

Which Way Is Morning?

If your dog wakes up with the sun shining in his face, his doghouse must face east. How can anyone make a statement like that without even seeing the doghouse and be right?

It all has to do with the way our planet behaves. The world we live on spins like a top, making one complete turn every 24 hours. That sounds slow, but it really isn't—only the fastest jet planes and rockets can keep up with that.

Although we always talk about the sun "coming up" or "going down" in the course of a day, it's actually the spinning of the earth that makes the sun look as if it is moving across the sky. You can see for yourself how this works.

Answers: 423 Park Blvd., 7; 1075 E St., 1; 1156 C St., 9; 527 Peach Ave., 4; 340 Cypress Ave., 11; 1117 D St., 3.

21

The World in Your Hands

For this experiment, you'll need a flashlight or a small lamp, a large ball, and a place that is kind of dark. You can work on the floor, but it may be easier to see if you have a table that is at least as long as you are tall.

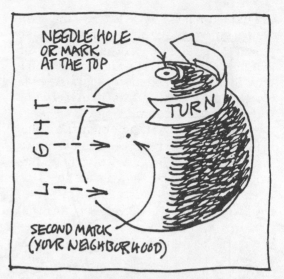

Put the ball at one end of the table (or at least five feet away from the light) and aim the light at it. You have just made a model of the earth (the ball) and the sun (the light).

If you walk over to the ball and look it over, you'll notice that only about half of it is lighted by the light. That's just the way things work with the sun and the earth—half the planet is dark, while on the other side it's daytime. It would not be much fun to live in either half if that's how things stayed, because the light side would get too hot, while the other side would freeze.

Luckily for us the earth is restless, turning around once every 24 hours. Practically every part gets some daylight and some darkness.

You can turn daylight into night by taking your world into your hands and giving it a spin. But first you need a couple of reference points on your ball. Your earth/ball may have a needle hole, or some other spot, already marked on it. If it does, turn the ball so that the mark is on the top. If not, make a mark there. Then make another mark—one that will come off easily might be best for you and the ball—about halfway down on the lighted side. This second mark represents your neighborhood.

If things were this way in reality, it would be around noon in your part of the world. But the world can't stand still, so yours has to get going too.

You can see what happens by placing one finger on the top spot and slowly spinning the ball from left to right. If you look down from the top, the ball should turn counterclockwise (opposite from the way clock hands move).

Notice that your neighborhood dot moves from the light to the edge (sunset), around the dark side (night), and back into the light on the other side (sunrise). Your dog is just getting the first rays on his face. To him it looks as if the sun is rising, but you now know that it's really just that the earth is spinning into the light.

NONSENSE! YOU CAN'T TEACH AN OLD DOG NEW TRICKS, AND I STILL SAY THAT IT'S THE SUN THAT RISES UP.

SURE, BOY. EVERYBODY ELSE CALLS IT THE "SUNRISE" TOO.

That mark at the top of the ball represents a place on earth called the North Pole. The rest of the planet spins around, but the North and South Poles (the South Pole is exactly opposite the North Pole) stay put as if there were fingers holding them in place.

Another way to think about the poles is to imagine the world as a marshmallow that is being roasted. The poles would be the places where the stick goes through the marshmallow. You could turn the marshmallow around, but the spots where the stick goes through stay where they are. Until the marshmallow melts, that is.

You probably know a thing or two about the North Pole. It's famous as the home of that jolly old fellow in the red suit who is supposed to bring presents to good children at Christmas. The North Pole is also famous for being colder than Groundhog Day in Duluth.

Here are some chilling things to think about. If the North Pole is so cold, what is it like at the South Pole? Would you need a coat there? Well, even your fur coat would need a coat at the South Pole, because it's just as cold as it is at the North Pole. We usually think that the weather gets warmer the farther

south we go, and that is true to a point, but the South Pole is far beyond that point.

The North and South Poles are the most important reference points on earth since they are the basis for our cardinal directions. Nope, not the red, crested grosbeaks you may see flying around your backyard, but the four main points we use to tell us which way we are headed.

From any point on earth, there is only one shortest way to get to the North Pole. When you turn that way, you are facing the direction we call north. If you face the shortest line to the South Pole, you are looking to the south.

So what's the big deal? Well, without a directional frame of reference, it would be really hard to tell someone else, or have them tell you, how to get from one place to another. It would work only if you both started in the same spot facing the same direction.

A treasure map is a good example of this. If the map showed that you should go 50 paces to a palm tree, you'd have to hope that there is only one palm tree. It

wouldn't help if the directions said "50 paces straight ahead," unless you happened to be facing the same way the mapmaker was when the instructions were written.

If the map said to go 50 paces south, you would know exactly which direction to face in heading toward the treasure. It wouldn't matter at all which way you were turned when you read the directions, as long as you knew or could figure out which way was south.

The same thing would be true for two people living on opposite sides of the street. Instructions telling each of them to turn right when they walked out their front doors would have them heading in opposite directions. If they were both told to head north, they would both go the same way.

So north and south are two of the cardinal directions—and they are always opposite each other. But are two directions enough? What if you want to show directions between north and south? Two more cardinal directions —east and west—serve as those in-betweens. They are opposite, too, although there are no east and west poles.

The easiest way to see how the whole cardinal system works is to use that scientific instrument you carry with you everywhere you go, your body. Assume for a moment that your nose is pointing north. That would mean that directly behind you is . . . that's right, south. Then if your right hand is east, your left would be . . .

Look, ma, a human compass!

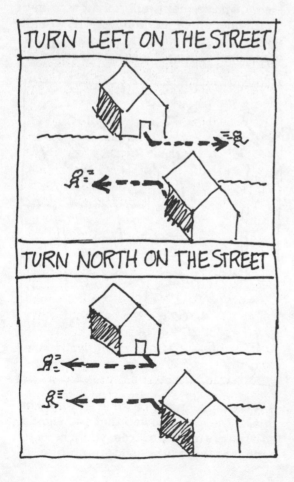

The Sun as Your Guide

The sun can serve as your guide in lining up the four cardinal directions in your neighborhood. Remember the dog with the morning sun for an alarm clock? Morning is a good time to look around and get your sense of direction because the sun rises in the east. Sunset works just as well since the sun always sets in the west.

By checking with the sun, you can tell a great deal about the way your neighborhood streets are lined up with the cardinal directions. Most towns are laid out so that the streets run in a north-south and east-west pattern, or very close to it.

Are your town's streets laid out north-south and east-west? If they are, is your street a north-south one or an east-west one? The way neighborhood streets are laid out may affect your life more than you realize.

It might have something to do with the reason you like walking home from school on one side of the street in January, and the other side in June. Have you ever noticed that parts of your house seem more inviting at different times of the day, and at different times of the year?

SEE IF YOU CAN FIGURE OUT WHAT THE SUN HAS TO DO WITH THESE TWO SITUATIONS.

Frank and his friends spent a lot of hot afternoons last summer playing basketball in the lot just east of the four-story parking garage. Now that it's October they play there only in the mornings and say that it is too cold in the afternoons. Why?

Sally loves baseball and plays the outfield. She can play left field without

any problems, but she likes right field better. The trouble is that she has been missing fly balls in right field when she plays that position in the morning, although she does well there in the afternoon. What's going on?

The reason Frank and his friends like the lot so much on summer afternoons is the same reason they don't like it on fall and winter afternoons. It's in the shadow of the tall parking garage to the west. That makes the lot cooler on any afternoon—a welcome relief in the summer, but too cold in the winter.

The reason Sally has trouble with fly balls in right field has to do with the direction in which the baseball diamond is laid out. When the right fielder faces home plate, she faces east, so Sally is looking into the sun during the early morning hours. That's why she has trouble following fly balls then. In the afternoon the sun starts to sink behind her, so it doesn't bother her at all then. Why do you think she never has trouble with the sun when she plays in left field? That's correct, the left fielder faces more to the north when looking toward home plate.

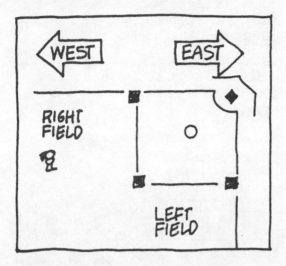

The streets in your neighborhood may not be laid out in the usual north-south, east-west grid. Sometimes lakes, rivers, mountains, or hills make it too difficult to run the streets along the cardinal directions. Or maybe the people just got tired of seeing everything so neat and regular.

Never fear, there are other directions besides the cardinals that have names, although they do borrow from the four main directions.

Something you can do to help you remember all of these directions is to make yourself a compass card. All you need for this is a sheet of paper, a pen or pencil, a pair of scissors, and something round.

Folded Directions

For this job you need a round piece of paper. Unless you can get a big round coffee filter from the kitchen, you may have to make the round paper yourself. If that's the case, find something round (a saucer, coffee can, or pot lid) that is small enough to fit on your piece of paper.

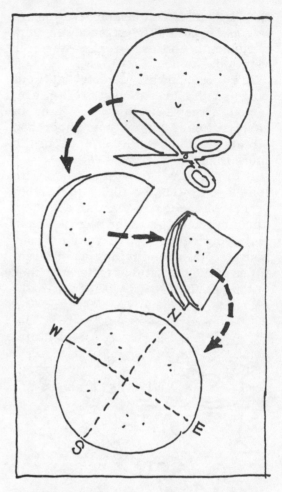

1. Trace the circle onto the paper and use the scissors to cut it out.

2. Now take the circle and fold it in half. You have just created a north-south line, and if you fold it in half once again, you will have an east-west one.

3. Open the paper and mark over the creases representing the four cardinal directions. You can mark them now with their initials — you know what they are.

Here's how to find the in-between directions. Fold up the paper again the way you had it folded in step 2, then fold it in half once more. Now when you open it up, you will notice that the new creases are halfway between the cardinals. Trace over the creases and you will have eight directions.

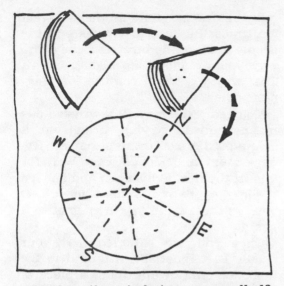

What are these in-betweeners called? Since they are halfway between the cardinals, their names combine the initials of the nearest cardinal directions. Can you guess what they are?

Starting from the north and working clockwise, the initials are NE, SE, SW, and NW.

CLOCKWISE IS THE WAY THE HANDS GO ON A CLOCK.

What about in between the in-betweens? Okay. Fold the paper back again the way you had it before, then fold it yet once more. It is getting pretty thick isn't it? When you unfold it this time, there will be eight new creases. Guess what directions those creases represent.

If you need a hint, they each have three initials. Another hint? The first initial of each is the closest cardinal. Got them?

The Magic Needle

Suppose you want to know which way is north, or some other direction, and you can't use the sun. That happens a lot—on rainy days, for instance, or at night. At those times, you need the magic of magnetism.

Unless you've been hiding under a rock all your life, you already know something about magnets. They are pieces of iron that have the power to attract other materials containing iron. People have known about natural magnets, which are called "lodestones," for thousands of years, but they have been used for only a few hundred years to help with direction finding.

How does a magnet find directions? One end of a magnet is attracted toward the north, and the other end toward the south. Knowing a magnet will behave in this way means it can be used as a direction finder.

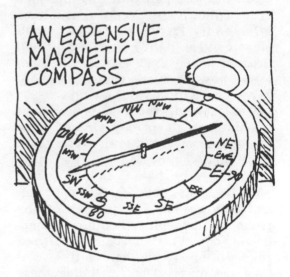

AN EXPENSIVE MAGNETIC COMPASS

If you have ever seen a magnetic compass, you know that it has a needle, a card with the cardinal directions printed on it underneath the needle, and a glass covering both of them. Compasses can be very fancy—and expensive—but you can make one that works with things you have at home.

Make Your Own

First you need a bowl or a bucket with water in it to be your compass case. It has to be made out of a material that doesn't contain iron, so avoid metal containers. Glass or plastic should work well.

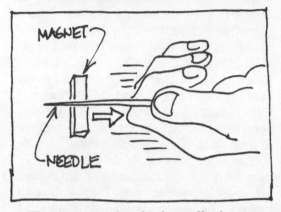

Then you need a steel needle for your pointer. Sewing needles seem to work best, and you can probably find one around your house. You will also have to have a float for your needle—a small piece of very light material that will float on water. Cork and plastic foam work best, and a small piece of cardboard will float for a while.

The last thing you need is a magnet of some sort. You may already have one of your own, but even if you don't, there may be dozens of these little things in your house. Cabinet doors often have magnetic catches on them, and they are used in just about any potholder, noteholder, or other strange-looking object that can be stuck to your refrigerator. A number of tools and toys are magnetized, including those stick-on plastic letters that your little brother or sister plays with.

You don't actually have to use the magnet in your compass—you only need it to magnetize your needle. You won't have to move the refrigerator! Simply pull your needle toward you lengthwise across the surface of the

magnet. Repeat that motion at least ten times. Be sure to always pull the needle in the same direction, the same way you would stroke a dog or cat. That way you won't confuse the magnetic poles.

You can test how well your needle is magnetized by trying it out on a nonmagnetic iron or steel object. Refrigerators and stoves work well for this. If the needle holds on and is hard to pick up, it's magnetized enough. If you can't feel a pull, try to magnetize it again.

Once you are satisfied that your needle is magnetized, attach it to the float. You may want to use a piece of tape to keep the needle in place so you don't have to go fishing for it later. Now you're ready to launch your magnetic voyager!

Hints: A couple of things may interfere with your magnetic marvel. One is any large piece of metal that's nearby. If you place your compass next to a metal pole, it will point to that pole, and *not* the North Pole. The other thing that could upset the directional apple cart is the size of your container. If the container is too small—or the float too large—the float and needle

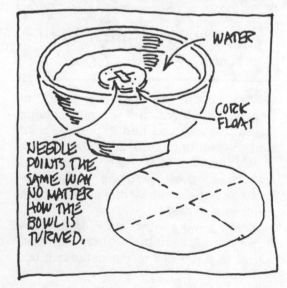

28

will be drawn to the side. You won't get an accurate reading this way, so find a larger container.

Once your float has had a chance to settle down, notice the direction in which the needle is pointing. If you turn the float around, what happens? Now try turning the container around (carefully—no tidal waves). What happens to the needle?

Take a look around at the way the needle lines up with your street, and the cross street nearest your house. If you have already figured out north and south for your neighborhood, how does the needle line up with those directions? You can take out the compass card you made earlier and lay it out the way you think it should be. Remember that the needle should line up in a north-south direction, but which end is which? Once you figure out that, you will have a pretty good idea of your neighborhood's layout. The little device you just made, simple as it is, has helped mariners find their way across uncharted oceans and assisted desert caravans in finding the next oasis.

The Neighborhood Mapmaker

Now that you know so much about your neighborhood, it is time to put some of that good old knowledge together to make a map. With a map you will be able to see in one place all of the various things you know about the streets where you live. You'll need a large sheet of paper, a pencil with a good eraser, a ruler (or some other straightedge), and a little patience.

How do you begin making a map? If you start with the outer edge, you will have a good idea of just how much of an area can be included within. But you may have trouble getting things to fit together if you only work from the outside toward the middle.

If you start in the middle and work toward the outside, the central part of the map will be right, but you won't know just how much you can fit on the sheet.

29

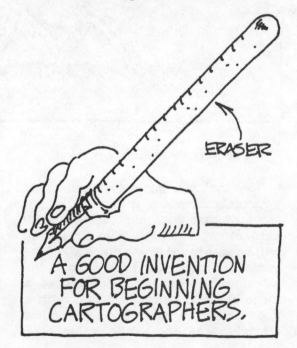

GET A BIG PIECE OF PAPER FOR YOUR MAP.

A way you can get out of this dilemma is to work on both—figure what you want to be the outer limits of your map *and* begin filling in some of the middle. Your mapmaking will be easier if your street pattern is a regular grid than if the streets go off in all directions.

Don't be afraid to use your eraser. Most beginning "cartographers"— that's what mapmakers are called— find that their erasers wear out faster than their pencils. Go ahead and map your neighborhood, no matter how you do it.

Try It Out

The real test of a map isn't how pretty it looks—getting it right is more important. The best way to check your map's accuracy is to take it for a walk.

See if you have forgotten major features such as streets, and check to see if they are in the right places. Did you get the names right when you labeled them?

Another very important thing to check is scale. The scale you need to be concerned with isn't the weight of the map, it's the way the size of things on your map looks compared to the way things look in reality. If two streets are about the same length, they should be about the same length on your map, not one twice as long as the other.

ERASER

A GOOD INVENTION FOR BEGINNING CARTOGRAPHERS.

Getting an accurate scale is about the toughest thing for a mapmaker to do. When places and streets are drawn to scale, the pieces fit together much better, and people using the map have a much clearer picture of the way the neighborhood really looks. (You might want to look at page 13 for a review of how scale works.

Is that the sound of an eraser?

A Note About Detail

If you find yourself using a magnifying glass when you write the names of streets on your map or putting one letter on top of another, you may have included too much.

You have shrunk your neighborhood down pretty far—putting the whole thing onto one piece of paper—so you don't have room for everything. Just how many things you can include (details) depends on the size of your paper. The larger it is, the more you can squeeze in. You can use abbreviations and symbols to put in some details and save space.

Don't forget to put a small compass with the cardinal directions on your map. Put it someplace where it won't interfere with things you want to show.

CHAPTER THREE
All Around the Town

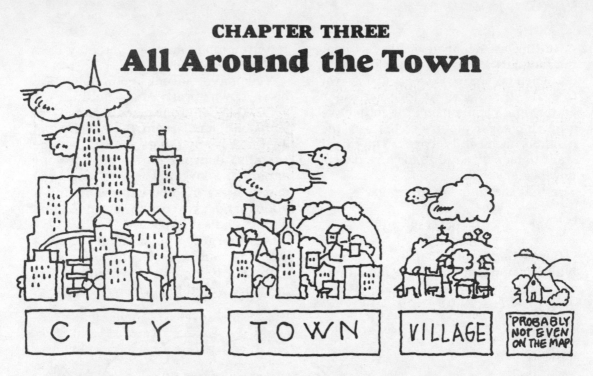

CITY　　　TOWN　　　VILLAGE　　"PROBABLY NOT EVEN ON THE MAP"

Most people live in a city or town, or at least very close to one. That city or town is home to them, just like their neighborhoods and houses are. How much do you know about the town or city that you call home?

For that matter, what is a city and what is a town and what are the differences between them? Both are places where groups of people live and work. The people who live there depend on each other to do the individual jobs that keep things going from day to day, instead of doing everything for themselves.

The main difference between cities and towns is mostly a matter of numbers. When you have thousands of people jammed into a square mile of space, you have a city. Towns are usually considered to have fewer people and less crowded conditions, but even large cities are sometimes called towns, and some small towns think of themselves as cities. No matter how big they are today, they all started out as small villages that could hardly even be called towns.

Here you are going to get a chance to take a closer look at the city or town where you live and get to know it better. This may be a little confusing for you if you live on the outskirts of a city in one of the surrounding communities called suburbs. You live in a town with its own name, but it is really part of the larger city. For this time at least, you have to think bigger.

A TYPICAL BEDROOM COMMUNITY

Say, What Kinda Place Is This?

People do like to compare things. There are whole books devoted to telling us just who ate the biggest doughnut, which word contains the most letters, how much the heaviest wombat weighs, and where the tallest treehouse is located.

Your town may not qualify as the biggest, most, heaviest, or tallest of anything, but it is unique in some ways. Do you know what makes your hometown special?

If you have traveled around to other places, you may already have some ideas about the differences. But if you haven't thought about it much, you will want to do a little research on your own. Adults make good subjects for a survey, since most of them think they have really been around.

Find a good corner (one where many people pass by) and ask people if they can think of two or three things that are outstanding about your city or town. These may not be good things necessarily—some towns are unusually smelly or noisy, for example. If you find that some things are men-

tioned by several people, you have probably stumbled across the characteristics that are considered special.

You may be more interested in facts than just opinions, and these you can find for yourself. What is the population of your city or town? Do you live in or near one of the 15 largest cities in the United States? One-fourth of the people in this country do.

What is its elevation above sea level? Or is it below sea level? What are its major industries and points of interest for visitors? If you check these facts against those of other cities, you'll have a good idea of how your town compares.

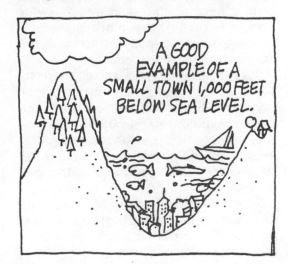

Get a Map to
Get the Picture

A good way to get to know your town better is to get a map of it. A map is as close as you can get to having a giant picture, and it has a lot of useful information besides.

There are many different kinds of maps of your hometown. Maps are kind of like trucks—they come in different sizes and shapes, and each has a particular use.

For example, there are maps showing all the electrical lines in your town. Others show the underground water pipes. There are even maps of the city sewer system. If you live in a city with bus service, there is a map of the routes buses travel, and probably a map showing the postal delivery (ZIP code) zones.

Any of these could tell you some things about the town you live in, but you would find a street map more useful. Aren't you more likely to walk or ride along city streets than through the sewer system?

Getting your hands on a street map of your city shouldn't be too difficult. Your parents may already have such a map, and if they don't, you can bug them to get one. Free street maps are getting hard to find, and adults—who always seem to have more money to spend than most kids—often have an easier time getting them.

If your parents don't know where to get a map, you might suggest gas stations, visitor centers, or the local chamber of commerce. Of course, you may want to use your own suggestions if your parents don't come through or have other ideas about who should get the map.

No matter how you get one, you will find the map your best guide for getting the overall picture of where things are in your town.

"Millie, you gotta help me," said Zeke to another cab driver and friend one afternoon in the Drowning Doughnut Diner. "The boss down at the Turkey Taxi office called me in today and told me that he's gonna fire me if I get lost one more time. What'll I do?"

"Relax, pal," Millie answered. "Like I told you before, what you need is a good street map. You won't get lost anymore once you learn to read it."

The two taxi drivers left the cafe and headed for Millie's cab. She handed Zeke an extra map she always carried with her, and he started unfolding it.

"Wow, will you look at all those lines and those little teensy words all over! I'll never understand how to use this thing," moaned Zeke. "And what are all these arrows and red dots and squares? And that green patch there? And that pink one over here? I should have stayed in bed today."

"Take it easy, Zeke, this thing isn't as bad as it looks; you just have to take the time to see how it all works. Most of what you see there, the lines and the

names, represent streets. Every single street in this city that has a name is on it."

"Well, what are those little arrows and all those other things? I don't know what they mean."

"Look over here in the corner. See that box that says 'Legend'? That's the place where you can find out what they mean."

Zeke looked at the legend and found all the funny markings there with an explanation of what they meant. The arrows showed which streets were one-way, and pointed in the direction traffic is supposed to go. The red dots on this map stood for hotels, and the red squares for public buildings. There was even an explanation for the pink and green patches — the green represented parks and golf courses and the pink showed cemeteries, hospitals, and other large public lands.

"Here, let's see how it works," said Millie. "We're right here at the diner on 25th Street," she pointed. Millie started the cab and took Zeke for a ride. He followed her every turn with his finger, tracing their route through the city. Sure enough, when she turned down Pine Street, it was one-way in the direction that the arrow on the map showed.

Zeke saw the city library whiz past just where the map showed it to be, and he was even able to tell Millie that the Van Snoot Hotel would be on the next block — just by looking at the map. For the first time since he had started driving his cab, Zeke knew exactly where he was, what streets would be next, and which way to turn to get where he was going.

He was so excited that he forgot to thank Millie for the map when she let him out next to his cab. "I'm never gonna be lost in this town again," said Zeke, turning on his radio. "The boss will be proud of me now."

Five minutes later, however, as Millie was letting a passenger out of her cab, she saw Zeke's cab pull up behind her.

"Millie, now I'm really in trouble. I just got a call to pick up a fare at 3150 Bluejay Way, and I don't have any idea where that street is. I don't even know what part of town it's in. And I thought this map would really help me."

"Calm down, partner," Millie chuckled. "I guess I forgot to tell you about the index."

"Index, appendix, I'm in trouble, Millie. If I can't find places I don't know, this is one Turkey Taxi driver whose goose is cooked," he wailed. "So what if I know where I am!"

Millie calmly unfolded Zeke's map and pointed to one of the lower corners. "See, the index lists every street in this town and the outlying areas. All you have to do is know your alphabet, because they are listed alphabetically. Here, find that street you mentioned."

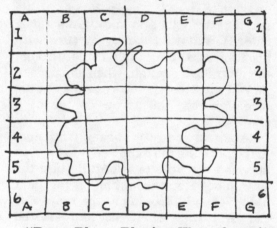

"B . . . Bl . . . Bluejay Way — here it is. But all it says is 'K-9.' What's that supposed to mean?" whined Zeke.

"Look over here at the edges of the map, Zeke. See these numbers running up and down, and see these letters running across the top and bottom edges? These little blue lines separate them and make a grid on the map. If you line up the "K" space with the "9" space, the block in the grid where they come together is where you'll find Bluejay Way."

Zeke found the K and, with his finger as a pointer, followed the light blue lines horizontally across the page. With his other hand he traced the 9 space down until his fingers touched. The block he had found was K-9, and, sure enough, there was Bluejay Way.

"Hey, I'm off to pick up my passenger," Zeke yelled as he jumped into his taxi. "I never thought using maps could be so simple."

Zeke did learn how to use his map, and he kept his job. In fact, he got so carried away with using his map that Millie had to remind him not to look at it while he was driving.

Zeke caught on pretty fast, once he had a little help. Take a look at your own street map and see what information it gives you. Do you have a street index and a legend (it might be called an "explanation" or "key")? What about the letter and number grid?

Now try finding these places — using the index or just looking for them on the map itself.

library	airport
police department	post office
railroad station	high school
city hall	cemetery

See how good you are at working the system backwards. Find these places, figure their coordinates (letter and number), then check the index to see if you are right.

park   sports stadium
church   bus station
golf course   college
hospital   courthouse

Another thing to check is the compass direction on the map. Practically all maps have a compass indicator somewhere that will tell you which way is north, and maybe the other three cardinal directions (south, east, west).

How are the streets laid out in your city? Is the grid pattern in line with the cardinal directions for the most part? If you made a map of your neighborhood, you will probably want to check it now against this street map.

Looking for Zeros

One thing that will help you know your way around your town better is to find the zeros.

Street numbering (and in many cases even naming) has a pattern that begins at a zero point somewhere in your town. If you are a good enough detective, you'll be able to find that spot and see how the pieces of the number puzzle fit together for the whole city.

Here are some clues that will help you find the zeros in your town. If you have a street map, you may be able to

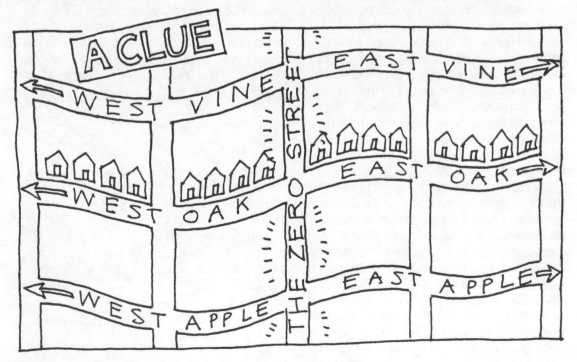

unravel the mysteries of the system right there at home.

Names. Sometimes streets change their names when they cross one particular street. West Oak Street suddenly becomes East Oak Street. If you see this happening to all the streets on the map as they cross the same street, you have most likely found one of the zero streets. The street numbers should get higher in either direction along the streets running away from this base, or zero, street.

The next trick is to find the other zero street, one that crosses the one you have just found. It is perpendicular to that base street and should be the place where the names change. Aha! Whipple Street just changed from North Whipple to South Whipple? That's the street that is the other zero, or base, street.

The intersection where the two base streets come together is your zero point. It should be the place in town with the lowest street numbers, no matter which way you go.

Downtown. The zero point is usually located downtown. That would be the oldest part of the city. It is actually called "downtown" because it has the lowest numbers. On a map, the downtown area of a city usually looks more crowded with streets, public buildings, and other points of interest than any other part of town. A major intersection there is a good candidate for the zero point.

Numbers. The most direct way to find the zero point, if not usually the easiest, is to follow the street numbers. Your map may be one of those super ones that has the hundred-number of the blocks printed on it. If so, it may actually give you the zero point, or show the zero/base streets at least.

If not, you might want to check out the system for yourself by taking a trip downtown. Hint: Not all street ad-

dresses are included in the citywide numbering system—especially short streets in the outlying areas. So even if your street is part of the citywide system and you have a very low number, you may be a long way from the zero point.

Check the cross street nearest you and see what the numbers are there. If they're something like 7800, it might be a long way to the zero point.

Once you know where the zero point of your town is, you will have a much better idea of where addresses can be found in the city. You might even be able to predict in what part of town 1265 West Aardvark Place can be found, even if you have never heard of it before.

38

TAKING TURNS

With these three turning tricks, show your friends that you can take turns better than anyone else. You can use the sidewalk and a piece of chalk or a good marking stone to show them how.

NO LEFT TURNS

A driver comes to the street where she wants to turn left, but there is a "No Left Turn" sign there. How can she get headed the right way on her street without making any left turns?

Answer: Instead of turning left (and maybe getting a ticket), she can go past the street and turn right on the next one. If she turns right at the next two corners after that and goes one more block, she will be headed the right way on her street. (Moral: Two wrongs may not make a right, but three rights do make a left, minus one block.) You might try giving this bit of back-seat wisdom to your parents sometime when they see the "No Left Turn" sign. Beats a ticket.

FIGURE EIGHT

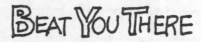

Willie Wanderer has decided that he is going to take an eight-block walk. He leaves the corner, which is his starting place, and goes one block north, then he turns west and goes two blocks, then north again for one block, then east one block, then he turns south and goes two blocks, then finally he goes east for one block. Are those his eight blocks? How many blocks away from his starting corner is Willie?

BEAT YOU THERE

Five friends have been planning to walk the six blocks to the skating rink as soon as school is out. They want to get there in a hurry, but they can't decide on which way to go. One wants to take a route that requires making only one turn, while another argues for a zigzag course that takes five turns.

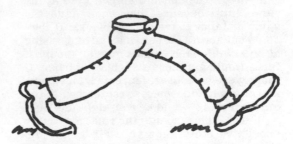

The friends each go separate ways, making one, two, three, four, and five turns in order to get to the rink six blocks away. If they all walk at the same speed, who will get there first? Why? How many streets will each of them have to cross on the way? (If this problem seems too tough, you can draw a map of the streets—it will look like a tic-tac-toe game with a border—but don't put in the routes each friend took until later. Then you'll look like a genius.)

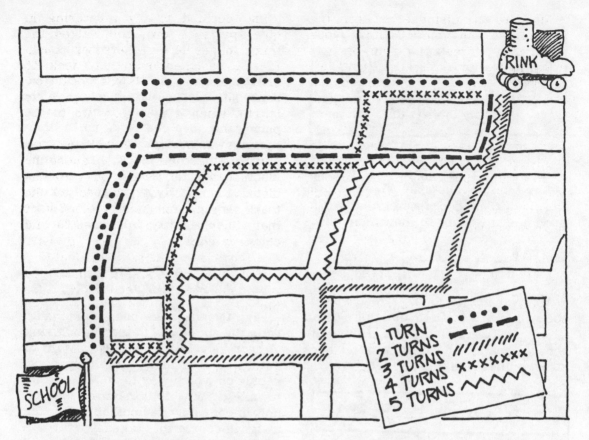

Answer: They should all get there at the same time because they each have the same distance to travel: six blocks. If they could cut across corners instead of crossing at crosswalks, the friends making the most turns would get there a slight bit earlier. If there were no houses, fences, vicious dogs, and angry homeowners to contend with, the quickest route would be a straight diagonal path from the school to the rink. Strangely enough, no matter how you walk toward the rink, you have to cross at least six streets.

How Far Is a Minute?

That question doesn't make sense. A minute is a measure of time. "How far?" asks for *distance*. Yet it is an important question for you to be able to answer for yourself. Then you'll know how long it takes you to travel a given distance, and just how far you can travel in a given amount of time. If you have that information, you won't have to be late for dinner—unless you plan to be.

First you need to know how long your pace is. Let's consider your pace to be the distance you travel each time you take two normal steps. The easiest way to find your own pace is to get a couple of friends to mark the places where the tip of your right shoe touches the ground. Or you could walk through a puddle or wet spot and use your footprints for markers.

40

Measure the distance between the two places and you will have your pace. Don't forget that your pace is actually two steps long, since your left foot also touched the ground in between the places your right foot touched. Once you have this reliable gauge of the ground you cover while walking, you can measure other distances. Like the length of your block, for instance.

How many paces long is your block? With the chart below you can figure out how long your block is in feet, if you have a consistent pace. It sure beats measuring it with a ruler or a yardstick!

Ellen used the chart to find out how long her block is. She measured her own pace at 52 inches and found that it took her 82 paces to walk her block. When she looked at the chart, she saw that a 52-inch pace at 80 paces per block meant that her block was about 347 feet long.

She decided to try measuring in other parts of town and walked the much longer block where her cousin Mark lives. Ellen found that it took 125 of her paces to walk Mark's block. How could she use the chart to figure its length, since it only goes up to 100 paces?

That's right, she added together numbers that total 125. Ellen added the distance for 50 paces (217) to the distance for 75 paces (325) to get the total distance for Mark's block (542 feet). She could also have used 60 and 65 paces since they also add up to 125 total paces, and the same 542 feet.

Mark was curious about what Ellen was doing and decided to try it for himself. He measured his pace at 46 inches. How many paces would he have to take to walk Ellen's block? Can you figure this? If you got an answer of around 90 paces, you have the method down pat.

YOUR PACE (OR MINE?)

Your Pace (inches)

Number of Paces	34	36	38	40	42	44	46	48	50	52	54	56	58	60	62
50	142	150	158	167	175	183	193	200	208	217	225	233	242	250	258
55	156	165	172	183	193	202	211	220	229	238	248	257	266	275	284
60	170	180	190	200	210	220	230	240	250	260	270	280	290	300	310
65	184	195	206	217	228	238	249	260	271	282	293	303	314	325	336
70	198	210	222	233	245	257	268	280	292	303	315	327	338	350	362
75	212	225	238	250	263	275	287	300	313	325	338	350	362	375	388
80	226	240	253	267	280	293	306	320	334	347	360	373	386	400	413
90	255	270	285	300	315	330	345	360	375	390	405	420	435	450	465
100	283	300	317	333	350	367	383	400	417	433	450	467	483	500	517

(All numbers given inside the box are feet.)

The Human Speedometer

Now that you know how long your block is you can add an element of time and be your own speedometer. A watch with a second hand will do the trick. Just see how many seconds it takes you to walk your block at a normal pace. Got it?

Ellen found that it took her a minute and 40 seconds (100 seconds) to walk the length of her block. At that rate how long do you think it will take her to walk to her friend Polly's house six blocks away? (The blocks are all about the same size in her neighborhood.)

Answer: 600 seconds, or 10 minutes, to walk the six blocks. The total distance is about 2,100 feet, or four-tenths of a mile.

An easy way for you to see how far you can travel in a given period of time is to see how many paces you take in a minute. Walking at a normal rate, count the number of paces you take in 60 seconds. Now look at the chart and see how far you have gone in that minute. That will be your speed in feet per minute.

Of course, you don't often hear people talk about speed in those terms. It's always miles per hour, or mph for short. Those are what you see on all the speed limit signs in your town.

42

Would you like to know how many miles per hour you are walking? It's easy if you have the figure for how many feet you travel in one minute. If you cover 88 feet in that minute, you are traveling at a rate of one mile per hour (1 mph). That means that it would take you one hour to walk a whole mile (5,280 feet!).

Here are other figures you can use for your walking speedometer:
110 feet = 1¼ mph; 132 feet = 1½ mph; 154 feet = 1¾ mph; 176 feet = 2 mph; 198 feet = 2¼ mph; 220 feet = 2½ mph; 264 feet = 3 mph; 308 feet = 3½ mph; 352 feet = 4 mph.

Although these speeds don't sound very fast compared with the speeds at which cars travel, walking four, or even three, miles per hour is really moving on. At 4 mph a person would click off a mile every 15 minutes.

Think about the athletes who can run a mile in four minutes (and there are a few who can). They are traveling at an average speed of 15 mph! The trouble is they can't keep that up for very many miles.

Why not see what your own record for the mile is by walking one yourself? Since you know your own pace, you can use the figures above to gauge your mile.

Hint: The easiest way to make the trip is to walk only about half the number of paces listed on the chart below, then head back the way you came. Otherwise you will end up walking *two* miles to get back to your starting point.

PACES TO THE MILE	
IF YOUR PACE IS 26 INCHES LONG, IT TAKES 2,433 OF THEM TO MAKE A MILE.	
YOUR PACE	PACES PER MILE
28"	2,262
30"	2,112
32"	1,980
34"	1,864
36"	1,760
38"	1,667
40"	1,584

Now you know how far a minute is and how many minutes are in a mile. If you want to check how accurate your pace is over a longer distance, you can look on your city map and find the mileage scale. By walking in a straight line on your trip, you can use a ruler to measure on the map the route you covered and see how closely your route matches the half-mile measure on the map. If it is very close, you have a good paceometer.

Town Traveler

Now that you have become an expert on your own town and know how to read maps and find places and can estimate how long it takes you to cover distances, it's time to take a trip. How you do this and where you go depend on what your town is like and where you want to go.

You may feel safer going with a grownup, but whether you go alone, with friends, or with your parents, *you* should be the one who plans the trip. Where you go depends on your own interests—there may be a museum, a skating rink, an aquarium, a zoo, or even a playground that you want to visit in another part of town.

This doesn't mean that you have to walk. You might want to take a bus, or maybe even a car. However you go, you can use what you've just learned to plan the route.

Here are some things you can do to sharpen up and check your around-town skills.

1. Estimate the distance to the place where you are going.

2. Estimate the time it will take you to get there and the time to return.

3. Estimate the number of blocks you will have to travel—are they all the same length?

CHAPTER FOUR
The Great State of ...

If you've ever been to the top of a very tall building or stood on a mountain peak, you know what it looks like to see the land around you stretching for miles into the distance. Perhaps you could even see faraway towns and cities, other counties, or other states.

What you couldn't see are the lines that mark the boundaries between those places, the invisible lines that we use to divide the "here" from the "there" on our planet.

Invisible lines may seem unimportant or silly, but they have played a very important part in our history. Wars have been fought to determine the location of some of those lines.

Folks in these parts call me the Borderline Kid. I reckon it all started when I was born in that cabin back in the Great Smoky Mountains right beneath Clingmans Dome. It sat right smack dab on the line between North Carolina and Tennessee, so I never knowed whether I wuz a native of the one state or the other.

My mammy and pap up and moved when I wuz just a sprout, but it wuz all the same. The town we lived in then wuz Bristol—half in Tennessee and half in Virginny. But then pap got restless, and the family moved on out west.

I growed up and went to school as fer as the third grade in a place called Texarkana. Yep, that's right, the darn place sat square on the border between Arkansas and Texas. Even the tax collector didn't know which state we wuz in. Maybe that's why pap liked livin' right on them lines so much—nobody could figger out who should tax him.

When I got to be 14, I left home and started to work as a cowpuncher (never could figger why they called us that—I never really punched a cow in my life, nor poked one either). Twenty years I spent on the range, a goin' back and forth from Texas through Okla-

homa, Kansas, Colorado, Wyoming, and Nebraska, herding cattle till my pants used to get up and hop into my saddle afore I even climbed outta my bedroll in the morning.

I wuz so tired of herding other people's cattle that me and a few of the boys decided one day to get into a new line of work. Instead of driving them Longhorns north into Kansas, we took 'em east into Arkansas, sold 'em, and kept the money ourselves. I guess you might say we became cattle rustlers.

We'd go out and steal cattle from some ranch in one state and drive 'em as quick as we could across the border into another, so the state lawmen a-chasin' us would have to let us be. We wuz outta their territory. That's why they called us the Borderline Bandits, and that's how I got my nickname. When things got too hot, me and the boys would head south of the border into Mexico for a while.

After a while, they caught onto our operation, and we had to disguise the cattle. Once we drove 150 head of Angus disguised as beavers from Colorado into Nebraska. Gettin' the tails to look right was hard enough, but you shoulda seen the trouble we had puttin' false teeth on them steers!

All that border stuff finally caught up with us though. Me and the boys had just brought a small herd of Longhorns up from Arkansas into Missouri disguised as armadillos when someone got wise to us. We made a run for it, and we didn't stop till we hit Kansas City, where we figgered we'd be safe. Well, it turns out that we wuz in Kansas City all right, but we wuz still in Missouri. How wuz we to know that Kansas City is in Missouri as well as in Kansas. It wuz an honest mistake for a bunch of rustlers to make.

The worst part of being caught wuz that everybody had it in for us. Me and the boys did time first in one state and then in another. We broke up rocks in so many prisons that we came to be called the Boulder-Breaking Boys.

Now I'm retired from the rustling business, but I'm still working the border. I work for one of those big casino gambling places here at Stateline, Nevada, driving buses full of people from their hotels on the Californy side of the line (gamblin' ain't legal in Californy) over into Nevada.

In a few years, when I get too old to drive these durn buses anymore, I'm gonna get me a little spread right smack in the middle of Nevada, as far away from the border as I can get. Some people say that all this living on the borderline is making me a little cracked.

STATE OF CONFUSION

The country we call the United States of America is made up of 50 individual states. Each one has its own territory marked by borderlines, its own laws and traditions and its own citizens.

If you live in the state of Indiana, you are a "Hoosier," while people of the neighboring state of Kentucky are "Kentuckians." It's easy to know when you have crossed the border between these two states because the wide Ohio River makes a physical barrier.

But what if you traveled eastward from Indiana into the state of Ohio? How would you know when you crossed the state line? The border between them is one of those imaginary lines that helped the Borderline Kid escape the law until he made his mistake. There is no physical boundary in this case, only a political one.

If that isn't confusing enough, states themselves are divided into smaller political units, usually called "counties." Some states have only a few (Delaware has 3 and tiny Rhode Island has 5), while others have 100 or more (Texas has 245, partner!). And counties are sometimes divided further into townships, precincts, or districts.

Let's take a look at a couple of maps of the state of Confusion, a state you won't find anywhere but in this book. It's a lot like those other states, though.

The first map shows mainly the physical features of Confusion. The borderlines and names of neighboring states are the only political information on the map, and the only man-

made features are the two reservoirs in the northern portion. This map doesn't tell you anything about where the people live but gives you some idea of what the terrain of the state looks like.

The borders follow some of the natural features (rivers in the southern and northeastern portions, mountain ridges in the northwest), but just as often are only based on invisible lines.

The second map contains the same basic outline and some of the same information, but a lot more has been added. This map is called a "cultural" one, since most of it deals with man-made features and the places where people live.

Using the two maps, see if you can find out: (1) how many counties there are in the state of Confusion, (2) which states it borders, (3) which is the longest river in the state, (4) where the highest point in Confusion is, (5) which city has the most people, (6) how many counties border five others, (7) which part of the state is the most mountainous, (8) how many towns in Poker County have 5,000 or more people, (9) which county borders only two other counties, (10) how far it is from the airport at Bearbite Ferry to the airport at Mudport.

Answers: 1. Nine; 2. Amnesia on the north, Paralysis to the west, and Lethargy on the south and east; 3. Bearbite River or the Bearbite-Marshy combination; 4. Way-up Mountain in Summit County is the highest in the state, Hogback is in Amnesia; 5. Rapid Falls is the largest you can be sure of, although Confusion City's population is impossible to judge; 6. Two: Flotin and Noway; 7. The western portion, especially the northwestern; 8. Four: Hamboro, Beque Bar, North Biggerville, and San Taclaus; 9. Only Delta County in the southeast corner, bordering Mudd and Porker; 10. About 110 miles by air (or as the crow flies).

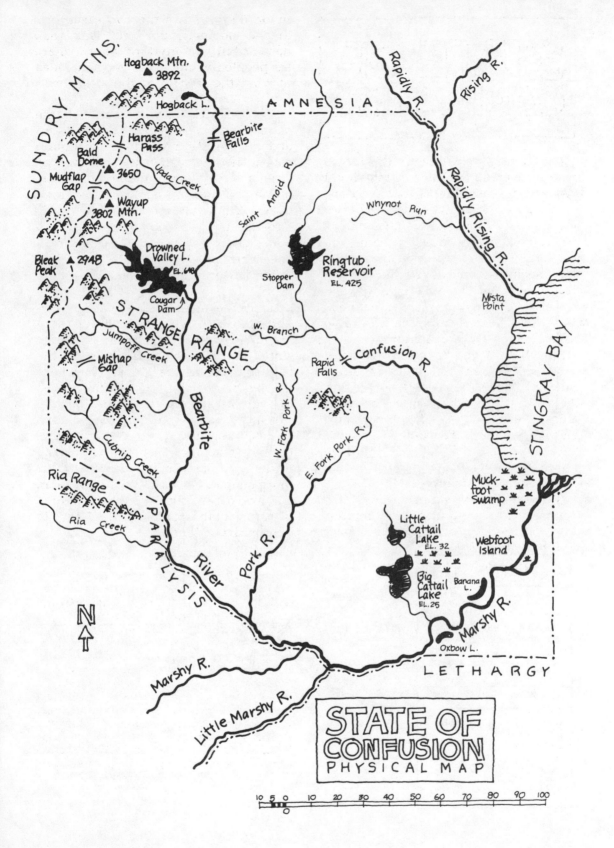

SUNDRY MTNS.

Hogback Mtn. ▲ 3892

Hogback L.

AMNESIA

Rapidly R.

Rising R.

Bearbite Falls

Harass Pass

Bald Dome

Mudflap Gap ▲ 3650

Ucla Creek

Saint Arnold R.

Whynot Run

Rapidly Rising R.

▲ 3802 Wayup Mtn.

Drowned Valley L. EL. 618

Stopper Dam

Ringtub Reservoir EL. 425

Mista Point

Bleak Peak ▲ 2948

Cougar Dam

STRANGE

W. Branch

RANGE

STINGRAY BAY

Jumpoff Creek

Mishap Gap

Bearbite

Rapid Falls

Confusion R.

W. Fork Pork R.

E. Fork Pork R.

Culprit Creek

Muck-foot Swamp

Ria Range

PARALYSIS

Ria Creek

River

Pork R.

Little Cattail Lake EL. 32

Webfoot Island

Big Cattail Lake EL. 25

Banana L.

Marshy R.

N ↑

Marshy R.

Oxbow L.

LETHARGY

Little Marshy R.

STATE OF CONFUSION
PHYSICAL MAP

10 5 0 10 20 30 40 50 60 70 80 90 100
0

49

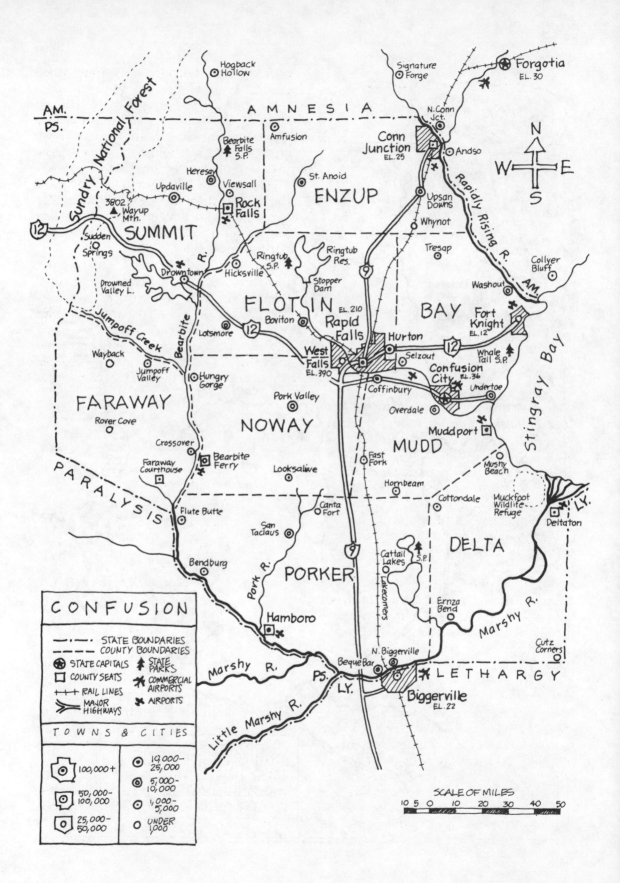

If you've ever looked at a map and asked yourself that question (or why is it yellow, green, or purple, for that matter), you already know something about the whims of mapmakers. Often the choice of colors simply shows clearly where one state ends and another begins—and no more.

The color of the states on maps has *nothing* to do with what's actually in the state.

Find a map of the United States that has well-marked boundaries between the states and see if you can solve these riddles of place. Where can you: be in the two states that border eight others? stand in four states at the same time? find yourself in a state that touches only one other? be in states that are divided, with one part attached to other states? find a state that doesn't touch any other state or country? discover the four states that border only two others? find a state with four straight sides that touches seven others? be in a state that doesn't touch any others, only another country? stand in a state that is smack in the middle of the country that borders only four other states?

If you used a map with colored states, you should have had an easier time figuring out the answers to those questions than if you had to look hard just to find the borderlines.

Here are the answers to the riddles of the great American jigsaw puzzle. The two states that border eight others are Tennessee and Missouri, and they also border each other.

You could stand in four states at one time in only one place in the United States. That's at Four Corners, where the states of Utah, Arizona, New Mexico, and Colorado meet.

The state that touches only one other is Maine, in the northeastern corner of the country, and the state it borders is New Hampshire.

Both Virginia and Michigan have parts that are detached from the rest of the state, separated by large bodies of water. Michigan's "Upper Peninsula" is attached to Wisconsin, while the "Eastern Shore" of Virginia is joined to Maryland.

The state that has no common border with any other nor even with another country is the island state of Hawaii. This group of islands in the Pacific Ocean is so far from the rest of the

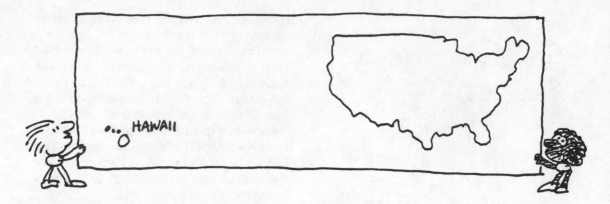

... HAWAII

states (2,000 miles!) that it usually appears only as an insert on most maps of the continental United States.

There are four states that share common borders with only two other states, but there are a couple others that might have tricked you. Washington (in the far northwest), Florida and South Carolina (in the southeast) are easy to spot, but little Rhode Island is harder to find. The two that might have fooled you are Delaware (which does border New Jersey, just across the Delaware River) and Michigan. Michigan is tricky because people sometimes forget the upper portion, which touches Wisconsin.

Two states have four straight sides, but only Colorado borders seven states.

Alaska is the only state that doesn't touch any others and is joined to another country. That country is Canada, the northern neighbor of the United States of America.

Even though it is smack in the middle of the country, Kansas borders only four other states. Its neighbors touch at least six others themselves.

I MAY BE A MEMBER OF THE UNITED STATES, BUT, FRANKLY, I'D NEVER WANT TO TOUCH ANY OF 'EM.

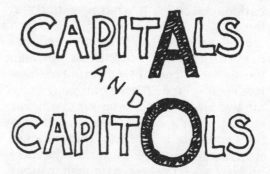

CAPITALS AND CAPITOLS

Hey, do you know the difference between a capital and a capitol? Did you know that you can have one without the other but not the other way around?

It's all very simple, really. A capital is the seat of government for a particular state or country, while a

A CAPITAL IS THE SEAT OF GOVERNMENT

capitol is a building where that government is actually seated. The capital of the United States is the city of Washington, D.C. The federal government is officially located there. And the U.S. capitol is there also. The Congress holds its meetings in that domed building. Where you find a capitol, you will find the capital city.

Confused? Well, things could be worse, and indeed they are. Many countries (including the United States) have political subdivisions that are called "states," "provinces," "districts," "regions," "departments," and "territories." These, too, have capital cities (with or without capitols) where the government of the area is located. In the United States all the states have both capital cities and capitol buildings.

Can you use a map to figure out which of these major cities are capitals of the states in which they are located? If they are not the capitals, which cities are?

Philadelphia	Dallas
Los Angeles	New York
Denver	Detroit
St. Louis	Seattle
Miami	Baltimore
Chicago	Minneapolis

Believe it or not, only one of these large, well-known cities is a state capital. If you guessed it was Denver (the capital of Colorado), you are right.

Even more remarkable is the fact that each of these cities is the largest in its state, showing that a capital isn't necessarily the biggest city. None of the five largest cities in the United States is a state capital, although two of them once served as the capital of the country in its early days.

Philadelphia isn't the capital of Pennsylvania, although it is many times larger than Harrisburg, which is the capital. Neither is Los Angeles, the giant city in California. Sacramento is the capital of California.

What about New York? Even though the city has the same name as the state and nearly half the state's population, the capital is the much smaller city of Albany, located 150 miles north on the Hudson River. Now use your map to find the others.

P.S. THE TWO CITIES THAT WERE THE U.S. CAPITAL AT ONE TIME ARE NEW YORK AND PHILADELPHIA. WASHINGTON, D.C., WAS MOSTLY A SWAMP THEN!

COMING SOON! CAPITOL HILL

53

What's in a Name?

The names given to many of the states in the United States are unusual. They are often hard to spell and even more difficult to pronounce, unless people are familiar with them.

Where did names such as Mississippi, Connecticut, Wyoming, and Massachusetts come from originally?

Most of them are Indian names, adopted by white settlers from the Native Americans who had been living there long before the first European ships set sail for the New World. In fact, 27 of the 50 states have names that can be traced back to Indian origins, as well as one (Hawaii) that was taken from the native language of the Pacific Islanders.

The names of six of the states come from the name of the group or tribe of Indians living in the area (Arkansas and Utah, for example), and others have interesting meanings when translated into English. Kentucky is supposed to mean "land of tomorrow" in the tongue of the Wyandot Indians, while Connecticut is best translated as "beside the long tidal river."

Not all of the names are as grand as are the translations of Alaska ("great land"), Michigan ("great water"), and Ohio (just plain "great" in Iroquois). Massachusetts has only a partly grand name ("great hill, small place"), while Arizona's name has a very un-grand meaning ("small spring"). Maybe the most unusual of all is Alabama, which is said to mean "I clear the thicket."

A dozen states have names that came from English sources, nine of them names of royalty or important government figures of the colonial period. Six state names come from the Spanish language (Colorado means "colored red" and Nevada means "snow clad"), and three are of French origin.

That leaves a grand total of only one state that has a name not related to Native American, Pacific Island, English, Spanish, or French peoples. This one state name is the only one tied to American politics. Can you guess it?

Can you find these words in the names of other states?

1. The name of another state.
2. A writing implement.
3. Two women's names.
4. What a sick person is.
5. How lemons taste.
6. An alcoholic drink.
7. A written rule that people live by.
8. What you do with a question (two states).
9. Noah's big boat.
10. An undergarment that is worn by women.

Answers: 1. ArKANSAS has the name of Kansas in it; 2. You could write with a PENnsylvania; 3. Nevada has both EVA and ADA—and there are others, such as IDAho, MontANA, InDIANA, MARYland, North and South CAROLina, not to mention GEORGIA, VIRGINIA, and CAROLINA; 4. Sick people feel ILLinois, and when they're ill, you don't want to annoy (INOIS) them; 5. Lemons are usually MisSOURi; 6. Grownups put VirGINia in their martinis and gimlet cocktails; 7. Legal matters are found right in the middle of DeLAWare; 8. You often NebrASKa or AlASKa question; 9. Remember Noah's ARKansas? 10. A woman may wear a NeBRAska under a New Jersey.

CHAPTER FIVE
Getting the Lay of the Land

If you traveled by airplane across the United States of America, you would see a great variety of terrain.

You would leave the narrow coastal plain of the northeastern states, hop over the hilly Piedmont area, and cross above the tree-covered forests of the long Appalachian Mountain chain. After passing the mountains, your plane would take you into the gently rolling interior of America's greatest river system. You would see the Ohio River and its many tributaries carrying their plentiful waters westward toward the granddaddy of our rivers, the Mississippi.

Beyond the muddy old river, the land below you would first become more even, then begin to rise gradually to the west. You would fly over the Great Plains, hundreds of miles of grassland prairies in the heart of the country.

Eventually, you would leave the plains behind. Then the land below you would become steep and rugged as you fly over the jagged backbone of the continent, the Rocky Mountains. Some of the peaks would have snow on them even in the summer because they rise more than two and a half miles above sea level.

West of the Rockies the country is still rugged mountain ridges alternating with highland valleys. You would fly over the Great Basin region, a vast area of high and very dry lands cut here and there by churning rivers. To the north is the Great Salt Lake, to the south the Grand Canyon of the Colorado River.

Flying westward you would reach another towering mountain range, the Sierra Nevada. The ridges of the Sierras slope gradually to their western foothills, which in turn lead down to the long Sacramento-San Joaquin Valley. One last hop over the jumbled rocks of the Coast Range would bring your 3,000-mile trip to an end at the edge of the Pacific Ocean.

Of course, a quick trip like this would

give you only a glimpse of the variety of landscapes in the United States. There is still a 2,000-mile expanse of ocean to the southwest separating the West Coast from the tropical-island state of Hawaii, and it is nearly that far to the northwest, to the chilly, giant state of Alaska.

Home of the Yankees

Long before the first of the Pilgrims set foot on Plymouth Rock, the earliest Americans were enjoying life in the Northeast. It was then, as it is today, a fairly rugged land of mountains and hills, broken by occasional valleys and a few deep rivers.

When you look at the little corner of the United States we call the Northeast, it's easy to think that it might be unimportant. The nine states usually included in this group (starting at the southern end with Pennsylvania and New Jersey) account for less than a twentieth of the land. But they are home to more than a fifth of the people in this country.

The area was developed by white colonists into a major world trading center. Cities such as Philadelphia, Boston, and New York, with their good harbors, grew and became wealthy as shipping and shipbuilding ports. The people became known to the local Indians as "Yankees," a name that has stuck with them for several centuries.

Six of the northeastern states (New Hampshire, Maine, Vermont, Massachusetts, Rhode Island, and Connecticut) are often called New England. Maine especially has a lot of Indian names that twist the tongue and mind. In Piscataquis County you can find Chemquasabamticook and Pemadumcook lakes, the Allagash River and Mt. Katahdin, which is just 12 feet short of being a mile high. And the county seat is Dover-Foxcroft.

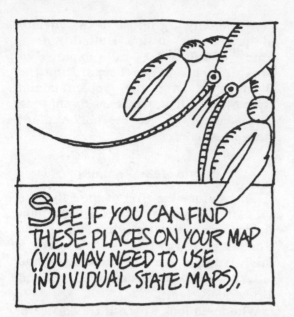

SEE IF YOU CAN FIND THESE PLACES ON YOUR MAP (YOU MAY NEED TO USE INDIVIDUAL STATE MAPS).

DOWN SOUTH

1. The city of Bangor, Maine, is on which river?

2. Part of the Vermont-New York border is formed by which lake?

3. Water from Lake Erie goes into Lake Ontario after it goes over a falls. What is the name of the falls?

4. What is the name of the river that separates New Jersey from Pennsylvania?

5. Which peak in New Hampshire is the highest in the entire Northeast?

6. The big hook of land that sticks out into the Atlantic Ocean in Massachusetts is called what?

7. The capital of New York State is connected with its largest city by which river?

8. Block Island is part of which state?

9. Most of one very large city in the Northeast is located on three islands. Which city is it?

10. Harrisburg is the capital of which state? It is located on which river?

Answers: 1. Penobscot River; 2. Lake Champlain; 3. Niagara Falls; 4. Delaware River; 5. Mt. Washington; 6. Cape Cod; 7. Hudson River; 8. Rhode Island; 9. New York; 10. Pennsylvania, Susquehanna River.

The southeastern states have their mountains too, but there a larger coastal plain and an area of rolling hills lie between them and the Atlantic. More level land and a warmer climate than in the Northeast made the Southeast an area of large cotton and tobacco plantations in the early years of white settlement.

Western parts of the region are in the Mississippi River basin. The "Big Muddy" snakes through its broad, low valley on its way to the Mississippi Delta, where it empties into the Gulf of Mexico. The gulf forms the southern boundary of the region and helps provide the warm, moist climate of the South.

THE BACKBONE OF THE APPALACHIANS

Looking at a map of the United States, you can see how the long Florida peninsula separates the Atlantic Ocean from the gulf, and how the Mississippi has built its delta into the gulf by bringing soil downstream for thousands of years. If you trace a line running roughly from Birmingham, Alabama, to Harrisburg, Pennsylvania, you will follow the chain of the Appalachian Mountains.

There are many rivers in the South since there is a good deal of rainfall throughout the year, but two are especially important. The Mississippi is a major highway of commerce, not only for the South, but for the entire central United States. Goods from as far away as Pennsylvania and Minnesota move downriver on barges to the port of New Orleans.

The Tennessee River and its many tributaries flow through seven states in the region, providing not only transportation but energy. How? The government-sponsored Tennessee Valley Authority (TVA) has constructed more than a dozen large dams that convert water power into electricity for the area.

The Southeast has many low, marshy areas near the coast. The Okefenokee, Dismal, and Big Cypress swamps are among the best known of these, as is the Everglades in southern Florida.

FROM THESE CLUES, CAN YOU FIGURE WHICH SOUTHEASTERN STATE YOU WOULD BE IN IF YOU:

1. Stood on the northeast bank of the Savannah River?

2. Were in a city called Memphis?

3. Dodged an alligator in Lake Okeechobee?

4. Waved to a steamboat at the mouth of the Tennessee River?

5. Sailed to the northern end of the Chesapeake Bay?

6. Visited a capital city named for an ocean?

7. Helped build a float for the New Orleans Mardi Gras celebration?

8. Tried to throw a dollar across the Rappahannock River, like George Washington did?

9. Found a capital that had the same name as the biggest seaport in South Carolina?

10. Had a picnic lunch along the banks of the Yazoo River?

11. Floated down a river through a capital named after a small stone?

12. Waited for a ship in a Gulf Coast city whose name means "able to be moved easily"?

13. Climbed the 6,684-foot Mt. Mitchell, the highest mountain east of the Mississippi?

14. Took pictures of the U.S. Capitol and the White House?

Answers: 1. South Carolina; 2. Tennessee; 3. Florida; 4. Kentucky; 5. Maryland; 6. Georgia (Atlanta was named after the Atlantic Ocean); 7. Louisiana; 8. Virginia; 9. West Virginia (Charleston); 10. Mississippi; 11. Arkansas (Little Rock); 12. Alabama (Mobile); 13. North Carolina; 14. You wouldn't be in one of the 50 states at all—you'd be in the District of Columbia, the federal area that includes just the city of Washington. We often hear the capital of the United States called Washington, D.C.

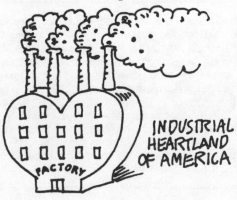

(map illustration showing CANADA, GREAT LAKES, ST. LAWRENCE R., USA, ATLANTIC)

In the Middle

North of the Ohio River and east of the Missouri is a group of states that used to be called the Northwest. But that was long ago, before the United States expanded all the way to the Pacific Ocean. Now these states are usually called the Midwest or, sometimes, the Great Lakes states.

The major features of this region are the Great Lakes to the north and the big rivers to the south and west. Most of the terrain is gently rolling or nearly level, with a few ranges of low hills. In fact, the land is so gentle that Illinois is known as the "Prairie State."

Those Great Lakes and rivers played a very important role in the development of the "Old Northwest" by providing good overwater transportation routes. With the opening of only a few canals, it became possible for goods and people to travel all the way from New York City to the lake ports of Cleveland, Detroit, Chicago, and even Duluth, Minnesota.

Transportation has been improved even more today with the completion of the St. Lawrence Seaway. Huge oceangoing ships can travel up the St. Lawrence River through Canada to any port on the five Great Lakes—a trip once possible only for small barges that were worked by mules. The St. Lawrence Seaway has made it easier to move the coal and iron found in the Midwest to the rest of the country and the world.

The Midwest has become the industrial heartland of America, as well as the center of transportation.

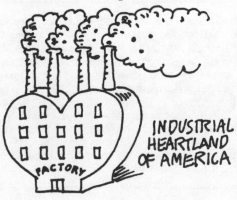

INDUSTRIAL HEARTLAND OF AMERICA

CAN YOU MATCH THE SCRAMBLED WORDS WITH THE CORRECT MEANINGS?

Bawash	Tonamines	Coinswins
Coria	Mousclub	Sadinilapino
Ripesour	Rotedit	Krazo
Chinamig	Sed Noisem	

1. Capital of the "Buckeye State."

2. Only Great Lake entirely within the United States.

3. City where the Ohio and Mississippi rivers meet.

4. River that partly divides Illinois from Indiana.

5. Madison is the capital of this state.

6. The capital of the state in which Sioux City is located.

7. Mountain range in southern Missouri.

8. Largest of the Great Lakes.

9. The capital that has the state name in its own name.

10. Iron-producing state with its twin capital and largest cities across the Mississippi River from each other.

11. Big Great Lakes port on the little Lake St. Clair.

NEWSRAS

Newsras: 1. Columbus, 2. Michigan, 3. Cairo, 4. Wabash, 5. Wisconsin, 6. Des Moines, 7. Ozark, 8. Superior, 9. Indianapolis, 10. Minnesota, 11. Detroit.

Where the Buffalo Roamed

Once there were so many buffalo on the Great Plains that travelers had to wait days for a herd to pass by. More than 50 million of the big animals roamed the grassy plains states less than a century and a half ago.

Nowadays, the former home of the shaggy American bison (buffalo) is where a large portion of the country's corn and wheat is grown, and the few surviving animals roam mostly in zoos. The vast grassland that stretches more than 1,200 miles from central Texas to the Canadian border and beyond has become America's granary.

The Great Plains slope gradually upward toward the Rockies in the West, reaching over a mile in elevation before the mountains begin. Rivers in the region flow south and east toward their meetings with the mighty Mississippi. Few trees are found in the plains states; those that are nearly all grow along the banks of these streams.

Besides the single row of states from North Dakota to Texas, the eastern portions of the states of Montana, Wyoming, Colorado, and New Mexico are part of the Great Plains region. And although these states are located in the geographical center of the country, some of the roughest names of the "Old West" are found there. Deadwood, Ft. Laramie, Dodge City, Wichita, Abilene, and Amarillo are places made famous by tales of cowboys, gold seekers, and outlaws.

The plains states are also famous as the home of powerful Indian tribes, such as the Cheyenne, Sioux, and Blackfoot, who were pushed off their lands by white soldiers and settlers.

Backbone of a Continent

The Rocky Mountains form the rugged spine of the North American continent. They are the tall divide that separates the rivers flowing east and south toward the Gulf of Mexico from those heading west to the Pacific Ocean.

In this range of mountains are dozens of peaks more than 14,000 feet high. In fact, the entire states of Wyoming, Utah, Colorado, New Mexi-

co, and Nevada have average elevations of over a mile high. So even if the Rockies weren't the backbone of the continent, they would be its roof!

Some of the most dramatic landscapes in the United States can be found in the Rocky Mountain states. From the great icy glaciers on the Montana-Canada border, through Yellowstone's steaming geysers, the Teton Range, the Great Salt Lake, through Royal Gorge, the Painted Desert, and the Grand Canyon of the Colorado River, nature displays some of its most spectacular scenery.

The Rockies are the birthplace of North America's greatest rivers too. The headwaters of the Missouri are high in the mountains of Montana and Wyoming, eventually combining with

THE ROCKIES ARE THE BIRTHPLACE OF GREAT RIVERS.

the Mississippi to form the third longest river system in the world. The Yellowstone, Platte, Arkansas, and Rio Grande are other important eastward-flowing rivers, while the Colorado, Snake, and Columbia are the principal waterways of the Far West that begin in the Rockies.

The region has long been important because of its valuable mineral resources. Gold and silver, copper and lead have all been found in parts of the region at one time or another. Today some of the wealth of the Rocky Mountain states comes from visitors to the mountains; they come to ski, hike, and fish, or just to enjoy the scenery.

ROCKY MTN. CROSSWORD PUZZLE

Across

1. The Sangre de _____ is a mountain range in New Mexico and Colorado.
4. Casper and Denver are located on forks of this river.
5. Devils Tower and the Teton Mountains are in this state.
9. The capital of New Mexico is _____ (also see #10).
10. Second part of #9.
11. Capital of the state east of Utah.

Down

1. Salt Lake City is the _____ital of Utah.
2. The Snake River forms part of its western border.
3. Four states meet at _____ spot in the United States.
6. City in the southwest corner of Arizona.
7. Largest Rocky Mountain state.
8. Yellowstone Park is famous for its steaming _____.
10. Second part of #9 across.

See page 67 for answers.

The Pacific Coast

At the western edge of the country are the Pacific Coast states. Although they have an ocean border like the Atlantic states, they are very different in their physical makeup. The coastline is rugged in most places, with mountains reaching right to the shore.

The climate is also very different from the East Coast. The southern half of California, with its great interior valleys, is very warm and dry—it's a desert in some areas—while the coast of northern California, Oregon, and Washington is very damp. Giant forests of redwood, fir, and spruce grow there in abundance, and the rivers are the spawning grounds for salmon and trout.

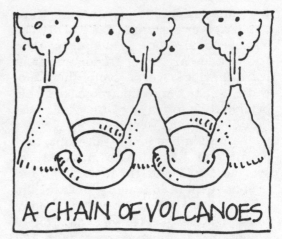

A CHAIN OF VOLCANOES

Perhaps even more impressive than the Pacific Coast are the mountains of the Far West. The Sierra Nevada of eastern California rise as high as the Rockies, although they are less than 200 miles from the ocean. The Cascade Range, which runs from California north to the Canadian border, is a chain of volcanic peaks.

Between these towering mountains and the sea are the fertile valleys of the West Coast states. These provide a large portion of the nation's food, including wheat, rice, apples, grapes, citrus fruit, and the greatest part of the nut supply. The mild climate of the region has attracted people from other parts of the country and the world, just as precious metals set off the first stampede of settlers in the middle of the last century.

The California gold rush actually started in 1848, but it took a while for the news to spread. So many fortune hunters crowded into the state the following year that they got the name "49ers," which has stuck ever since. For more than a decade, people scoured the foothills of the Sierra Nevada hoping to strike it rich. They left behind colorful town names such as Rough and Ready, Hangtown, Shingle Springs, Angels Camp, Mt. Bullion, and Dry Diggins.

IF YOU WERE A PILOT FLYING BY DIRECTIONS, COULD YOU FIND & NAME THESE PLACES?

1. A lake 80 miles ENE of California's capital city?

2. A narrow river gap 110 miles north of Bend, Oregon?

3. A body of water 140 miles due west of Spokane, Washington?

4. The snow-covered Cascade peak 110 miles northeast of Eureka, California?

6. The lowest point in the United States, 100 miles west of Las Vegas, Nevada?

6. Giant dam 120 miles NNE of Yakima, Washington?

7. The national park with huge sequoia trees, 150 miles east of Oakland, California?

8. Mouth of a great river, 250 miles NNW of Crater Lake, Oregon?

9. The bay 150 miles northwest of Fresno, California?

10. The West's largest city, 75 miles ESE of Santa Barbara, California?

11. An active volcano 75 miles south of Tacoma, Washington?

12. The city 330 miles WSW of Oregon's northeast corner and 110 miles SSW of Portland, Oregon?

13. The desert 150 miles north of San Diego, California?

14. A mountain over 14,000 feet high 145 miles NNE of Oregon's capital?

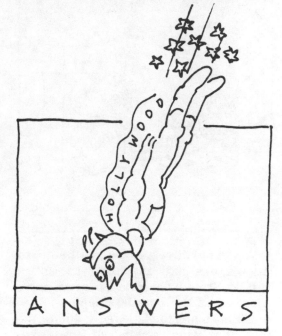

ANSWERS

1. Lake Tahoe, 2. The Dalles, 3. Puget Sound, 4. Mt. Shasta, 5. Death Valley, 6. Grand Coulee, 7. Yosemite, 8. Columbia, 9. San Francisco, 10. Los Angeles, 11. Mt. St. Helens, 12. Eugene, 13. Mojave, 14. Mt. Rainier.

Where East Meets West

Even if you had traveled the more than 4,000 miles from Key West, Florida, to the tip of Washington State's Olympic Peninsula, you would

THE LOWEST POINT

still have a long way to go before you found the farthest reaches of the U.S.A. You would have a trip ahead of over 2,500 miles—in two directions, in fact.

To the northwest is the giant state of Alaska, one-sixth the size of the entire United States. Despite its great size, Alaska has the smallest population of the 50 states. This is largely due to the fact that this northwest corner of North America is so far north that much of its interior is permanently frozen.

But the northern giant is also a land of bountiful resources. Timber and fishing are important industries along the wet and cool southern coast, where most of the population is found. Alaska appears to be rich in oil as well, with thousands of gallons traveling across the state from wells on the frosty Arctic Ocean coast.

Like California, Alaska first earned its reputation for its gold. Discoveries in the neighboring Klondike region of the Yukon (in Canada) and later the Seward Peninsula on Alaska's west coast brought Americans and other adventurers streaming into the far northwest. They found a land of great opportunity, but also a very rugged land that tested even the hardiest.

Alaska boasts the highest mountain in North America, 20,320-foot Mt. McKinley, and the mighty Yukon River. The Alaska Peninsula and Aleutian Islands are the home of a volcanic mountain range that has been very active in the recent past.

Volcanoes are about the only thing that Alaska has in common with its tropical sister state, Hawaii. That state is made up of islands that are tiny when compared with the U.S. main-

VOLCANOES ARE ABOUT THE ONLY THING ALASKA SHARES WITH HAWAII.

land—only the largest island, called Hawaii, is larger than little Rhode Island.

The Hawaiian Islands owe their very existence to volcanoes. Hawaii, the "big island," is one of the most active volcanic areas on earth, and gentle eruptions are still adding to its size. Mauna Kea and Mauna Loa are twin volcanic peaks there, each rising more than 13,000 feet above the sea. They are so tall that they occasionally get snow on top, even though the temperature 20 miles away rarely dips below 70 degrees Fahrenheit.

That tropical climate is what makes Hawaii one of the most popular tourist attractions in the world. And its central location in the Pacific Ocean has made it a place where the East meets the West.

ON WHICH ISLANDS WOULD YOU FIND EACH OF THESE?

1. An extinct volcano called Haleakala.
2. Waikiki Beach.
3. A town called Dutch Harbor.
4. Kilauea crater.
5. Sitka.
6. The town of Kaunakakai.
7. Pearl Harbor.
8. The town of Kodiak.
9. The rainiest spot on earth, Mt. Waialeale.
10. Katmai volcano.

ANSWERS

1. Haleakala is a 10,000-foot extinct volcano that you can find on the island of Maui in the Hawaiian Islands. 2. Waikiki is a famous beach in Honolulu on the island of Oahu, Hawaii. 3. Dutch Harbor is on Unalaska Island in the Aleutians, Alaska, in spite of the name. 4. Kilauea crater is an active volcano on the big island of Hawaii. 5. Sitka is a city on Baranof Island, Alaska. 6. Kaunakakai is the principal town on the island of Molokai, Hawaii. 7. Pearl Harbor is the giant naval base on Oahu, just west of Honolulu, Hawaii. 8. Kodiak is on the island of the same name. It is famous for its bears. 9. Mt. Waialeale is in the center of the island of Kauai, the fourth largest of the Hawaiian Islands. 10. The Katmai volcano isn't really on an island at all, unless you could consider the whole North American continent to be one. It's on the Alaska Peninsula for now, until a big eruption makes an island out of that peninsula.

Answers to crossword puzzle on page 63.

CHAPTER SIX
The Global Grapefruit

It has taken centuries for people to get a complete picture of the world in which we live. The idea that our home planet is a giant, spinning ball traveling through space at a speed of thousands of miles per hour isn't a conclusion you reach very quickly when all you've ever seen is a few square miles of that planet. But as people began to travel across increasingly larger portions of the earth, that's the picture that developed. Great oceans were crossed, new lands were discovered and explored, and the information gathered was recorded in stories and maps.

In this chapter we're going to look at some of the ways mapmakers have helped us understand and get around on this planet. There are two things you will need to help you: an atlas (a book with maps of the different parts of the world) and a grapefruit.

Grapefruit? Why not, as long as you are careful not to let your global citrus fruit squirt you in the eye. After all, there's no reason why geographical reading can't be done with good taste.

An Appealing World

Geographers have a difficult time when they have to look at the whole world at once. The most accurate way is to use a globe, a ball-shaped map that is round, like the earth. (Remember the night-and-day experiment on page 22?) But a globe doesn't give you a complete picture of the entire planet, since you see only half of it at a time.

For that reason (and because globes are hard to carry around and don't fit into books!) mapmakers have had to find ways of "peeling" the world so they can fit their pictures of it onto a single sheet of paper.

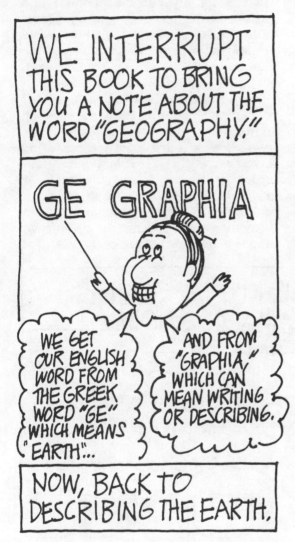

make a cut (just deep enough to go through the skin) halfway around the grapefruit to a spot opposite where you began.

This will give your fingers a place to start peeling the grapefruit. Carefully work your fingers around and under the edges of the skin as you pull it away from the fruit, keeping it in one piece if you can.

How did it come off for you? Not so easy, eh? Once you have the skin off, you are ready to make your map. Turn the yellow side up and flatten it as best you can. What kind of problems are

You can get some idea of how this works by trying it yourself. Get your grapefruit, a sharp knife, and some fingernails.

To make the peeling easier and to give your map straight sides, you want to make a shallow cut through the skin. Start with the knife at the stem end and

69

you having, and how does your world look compared to the way it did when it was still round?

Mapmakers have some of the same problems—minus the sticky fingers—when they try to make the curved surface of the earth fit on a flat piece of paper. You probably found that you need to make cuts in or tear the grapefruit skin in order to make it lie flat. But for a mapmaker this method presents a problem, since it leaves large gaps in the map.

To avoid having map gaps, mapmakers have discovered how to use mathematics to help them "stretch" the earth into a flat shape. If you were working with a hollow and very stretchable rubber ball, you could do this by pulling the top and bottom edges (after making that cut first) until you had stretched it into a rectangle.

That's exactly what Flemish cartographer Gerhard Kremer did more than 400 years ago. He developed a projection, a mathematical stretching method, that turned the spherical surface of the earth into a can shape, or cylinder. Then he simply cut the cylinder and flattened it to make a rectangle. Most of the maps you see of the whole world are Mercator projections, based on the method used by Kremer, whose cartographic pen name was Mercator.

Is your map of the world a Mercator projection? You might find the name in small print somewhere around the map's border, but even if you can't find it, you can still tell if it is a cylindrical projection by looking at the lines, or grid, on its face. If the lines form rectangles, it's a Mercator projection.

Another way to spot a cylindrical map is to look at the size of the island of

THE WORLD-FAMOUS MR. MERCATOR AND HIS FABULOUS MAGIC SHOW!

WATCH AS I CHANGE THIS GLOBE INTO A CYLINDER...

AND THE CYLINDER INTO A RECTANGLE.

Greenland. This island to the northeast of North America looks much larger on Mercator maps than it really is because of the stretching necessary at the top of cylindrical maps. It's actually less than a third the size of the United States—even excluding Alaska, which for the same reason looks much bigger than it really is.

There are other methods of making map projections, including stretching the world into a cone and slicing it as you did the grapefruit. Each of these has its uses but not one of them represents the earth without some distortion. And not one of them has the appealing taste of your grapefruit.

Halves

"Nations of the Western Hemisphere have sent representatives to an important economic summit meeting in Buenos Aires," the TV newsperson says. What is the Western Hemisphere, or any hemisphere, for that matter?

One way to visualize a hemisphere is to think of the earth as our old friend the grapefruit. This time we're going to let the stem end serve as the North Pole and the opposite end the South Pole. If you turned the grapefruit on its

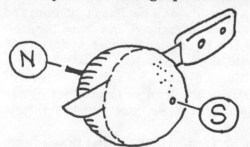

side and sliced it in half midway between the poles, you would be dividing it along the equator. This would give you two half-spheres, or hemispheres. The stem (top) half would be the northern hemisphere, and the bottom half the southern hemisphere. Earth's Southern Hemisphere

is sometimes called the "water hemisphere" since less than one-fifth of its surface is land. But the most important reason for making a distinction between the north and south halves of the earth is that they have reversed seasons.

January is a winter month in the Northern Hemisphere. Often the coldest weather of the year occurs then. At the same time, the Southern Hemisphere is having its hottest weather—January is a summer month there. Christmas in Buenos Aires, Argentina, and in Perth, Australia, might be one of the hottest days in those cities, and the Fourth of July one of the coldest, instead of the other way around, as it would be in the city of Kankakee, Illinois, in the Northern Hemisphere.

If you cut your grapefruit in half the other direction—through the stem and bottom ends—instead of around the equator, you could divide this model of the world into eastern and western hemispheres.

Most of the people in the Eastern Hemisphere (the continents of Europe, Asia, Africa, and Australia) didn't

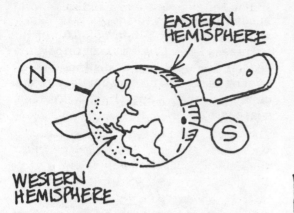

EASTERN HEMISPHERE

N

S

WESTERN HEMISPHERE

know of the existence of the lands of the Western Hemisphere until Christopher Columbus sailed across the Atlantic Ocean in 1492. The two American continents that make up the land masses of the Western Hemisphere were therefore referred to as the "New World."

CONVERSATION WITH COLUMBUS

Christopher Columbus (1451-1506) is the person most often credited with discovering the New World, and certainly his explorations opened the door for other Europeans to settle in North and South America.

Interviewed by *Discoverers Digest*, Columbus had some interesting things to say about those exciting days of the Age of Exploration. (He's had plenty of time to think about what happened because he made his famous voyages nearly 500 years ago.)

Discoverers Digest (*DD*): **Mr. Columbus, what first gave you the idea of attempting a voyage westward across an uncharted ocean at a time when most people still thought the world was flat?**

THE WORLD WAS FLAT

Christopher Columbus (*CC*): **Actually, I got the idea from some old Greek writings. Even in ancient times geographers knew that the world was round — a sphere — and not flat at all. Some of them even figured out how big the sphere is, so that's why I thought I could take a shortcut to the East by sailing west.**

DD: **You mean you were looking for a shortcut when you found the New World?**

CC: **That's right. At that time everyone was making shiploads of money by trading for silks and spices and all sorts of goods in India and the Spice Islands — what we called "the Indies." The only problem was that there were so many people who wanted a piece of the action. If you traveled overland through India, the Persians and Mon-**

gols would demand tribute; if you shipped by sea, the Arabs grabbed a big share. My idea was to cut them all out by heading west to reach the East. And it almost worked.

DD: Almost?

CC: Yes. My idea was okay, but I based my calculations on the work of some old guy in the Middle East who didn't have his head screwed on straight. He had estimated that the earth was much smaller than it really is, so instead of a 6,000-mile, easy sail west to the Spice Islands, I ran into the New World. You can't imagine how embarrassed I was when I had to tell Ferdinand and Isabella!

DD: You mean King Ferdinand and Queen Isabella of Spain?

CC: That's right.

DD: Weren't they happy that you had discovered a new continent, that you extended man's knowledge of the world?

CC: No, actually they were hopping mad. Well, not at first because it took everyone a while to figure out my mistake. When we landed, I thought we had hit one of the outer Indies, so I called the people there "Indians," and the name stuck, but I had really found the West Indies, not the Spice Islands.

Oh, the king and queen were happy enough when I came back from the first voyage. Isabella kept reminding Ferdinand that he had thought she was just throwing away her jewels when she had agreed to finance my trip. She was sharp though. She had heard that the Portuguese had found a shortcut to the East, around the southern tip of Africa, just a year before I made my first voyage west; and she was anxious to get a bigger share of the trade too.

But when I couldn't deliver anything more than a few Indians, some corn, and tobacco, they got mad.

SOME CORN AND TOBACCO

DD: But look at your achievements — the new lands you found and the way those discoveries changed the course of history . . .

CC: That's all very well for you to say now, but at the time of my death, I was in disgrace. I never got a chance to enjoy any commercial success from my discoveries while I was alive. And I never got the recognition I deserved after I died either. The continents of the New World were named after that

ANCHOVY-ONION PIZZA

THE QUEEN WAS HOPPING MAD.

Johnny-come-lately Amerigo Vespucci. I'm not the only explorer who made a mistake, you know.

DD: What do you mean?

CC: Well, take Ponce de Leon, for instance. He went off looking for the Fountain of Youth and found Miami Beach instead. And this guy Balboa — he climbed some hill in Panama, looked out, and named the biggest, meanest ocean in the world the Pacific, which means "peaceful."

LOOKING FOR THE FOUNTAIN OF YOUTH

And Cortes — he landed on the coast of Mexico with a handful of men and took on the whole Aztec nation. If it hadn't been for the fact that the Aztec king, Montezuma, believed Cortes was some kind of god, he and his men would have been the main course at a sacrificial dinner. Instead Cortes won the war and got the nation's gold.

DD: Did the search for gold have much to do with exploration?

CC: Of course. Don't think that explorers were sent out just for adventure or because of scientific curiosity! Why did De Soto spend three years wandering through the southeastern part of North America, or why did Coronado trek across the southwestern deserts? Explorers were searching for wealth in any form they might find —

WHATEVER WOULD FATTEN THE ROYAL TREASURY

gold, silver, precious stones, timber, animal hides — whatever would fatten the royal treasury.

The English hired my Italian countrymen Cabot and Verrazano to find a quick way to the riches of the East, while the Dutch hired Henry Hudson, an Englishman, to do the same for them. In the end most of us never got our fair share of the riches our discoveries made possible. We took most of the risks while the royalty of Europe got the wealth. If I had it to do over again, I'd stick to mapmaking.

IF I HAD IT TO DO OVER AGAIN, I'D STICK TO MAPMAKING.

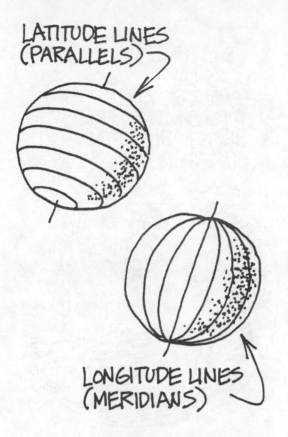

LATITUDE LINES (PARALLELS)

LONGITUDE LINES (MERIDIANS)

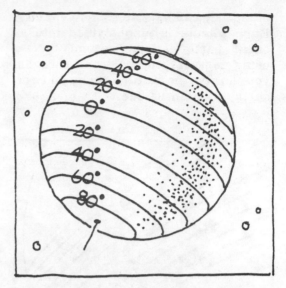

Zeros

Mapmakers have given the world a grid system something like Zeke's and Millie's maps. The grid lines running across the maps are called "latitude lines," or "parallels," and the up-and-down grid lines are called "longitude lines," or "meridians."

Remember the equator? It is the imaginary line drawn around the earth, much like a belt halfway between the North and South poles, and it runs in an east-west direction. The equator is the baseline for latitude—any point located on the equator has a latitude of zero degrees (0°).

Mapmakers have divided the areas between the equator and the poles into 90 degrees of latitude north and 90 degrees south. Another way to put it is that each of the poles has a latitude of 90°, one 90° north and one 90° south.

Places in between get numbers corresponding to their positions relative to the equator and the poles. (The lines are often called parallels because they all run in the same direction and are the same distance apart all the way around the earth.)

Take a look at a map of the continent of Africa and you can see how latitude lines work. Notice how the equator cuts across Africa from west to east and how other lines of latitude both north and south of the equator are parallel, running in the same direction.

Find Addis Ababa, the capital of the east African country of Ethiopia. You can see that it is located near the 10° north latitude line (written 10°N). Dakar, Senegal, at the western tip of Africa is located near the 15°N line, and Lusaka, Zambia, is close to 15°S. Your map may not have the 15° parallels, but even if it doesn't, you can see that both cities are about halfway between the 10° and 20° parallels.

For mapmakers, close is not always good enough, so each degree of latitude is further divided into 60 "minutes." (Yes, minutes, but here it means distance, not time.) That way, the latitudes of places between each degree of latitude can be stated more exactly. Each minute is also divided into 60 equal parts called "seconds." (By using degrees, minutes, and seconds, you can give the latitude of any place on earth to within 100 feet!)

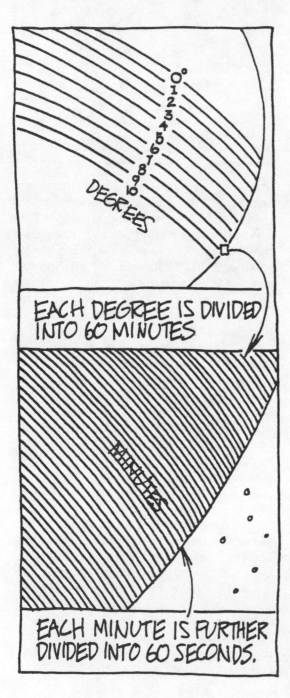

So the more precise latitude of Addis Ababa is 9°2'N (' means minutes). Dakar is at 14°34'N, and Lusaka is 15°28'S of the equator.

What about places that are east and west of each other? Cartographers (our mapmaking friends) have very thoughtfully provided us with a zero line similar to the equator for determining east-west positions of longitude. This imaginary line runs north and south from pole to pole, and meets the equator at the lonely zero-zero point in the Gulf of Guinea, 400 miles from the coast of Africa.

Although mapmakers surely chose the location of the zero longitude line for several reasons, we know it was first selected by English cartographers because it runs through the Royal Observatory in Greenwich, England. Then they drew the other lines of longitude

both east and west of the zero line until they met at 180° on the opposite side of the world.

Longitude lines are not parallel to each other, but are like the lines between wedges of a grapefruit. These lines come together at the poles—the only places on earth that aren't east or west of anywhere else. Another name for longitude lines is meridians, and the zero line is often called the "prime," or "Greenwich," meridian.

Take a look at Addis Ababa again. Notice that it is fairly close to the 40°E meridian. Since it isn't quite that far east, its actual longitude is less than 40° (38°42'E, to be exact).

Putting the latitude and longitude of the Ethiopian capital together, we get 9°2'N, 38°42'E. Once you know how to use these numbers —and you do now— you can pinpoint the location of any place on the globe. You won't need to know the *exact* location of places unless you are doing geographical

work, but the numbers and their directions give you a good idea of just what part of the world you are talking about.

Try to find the names of the places located at approximately 30°N, 31°E; 30°S, 31°E; 17°N, 3°W.

A HOLE TO CHINA

What about the place where you live? Can you figure out your own latitude and longitude?

What happens when you go west of 180°W? Then you have gone so far west that you have to start counting in reverse—180°W is the same as 180°E. It's also called the International Date Line—the time is one day later on the west side than on the east.

Have you ever heard the expression that someone is digging a hole so deep that the hole will go through to China? Have you ever wondered where you would end up if you *could* dig straight through the earth to the opposite side? Here's how you can figure the location of that spot. Whatever your latitude is, change its direction to the opposite one. (If you are at 42°10'N, make it 42°10'S.) The longitude is a bit trickier. First you have to subtract longitude from 180°, then change the direction of that difference to the direction opposite your longitude and you have it.

Example: For Honolulu, change the 21°20'N to its opposite (21°20'S). Then subtract 158°W from 180° and change the answer to the opposite direction (22°E). That makes Honolulu opposite 21°20'S, 22°E—a spot in the Kalahari Desert of southern Africa. For most of the United States the opposite side of the world is in the southern part of the Indian Ocean.

And what about China? If you are going to dig straight through the earth to that Asian country, you had better start digging in Argentina or Chile.

Time Out

There you are, sailing across the middle of the Pacific Ocean. You look at your calendar watch, and it tells you that the time is 9:20 A.M., June 15. Your ship is headed westward toward Japan.

Five minutes later, you glance at your watch again. It now reads 9:25, but you notice that something is strange. The watch tells you that it is now June 16! You shake the thing. Has it broken? Did its little gears go haywire?

Of course your watch won't really change the date all of a sudden, but if it could, this is the place where that would happen. It *is* a day later because you just traveled west across the International Date Line. This imaginary line is where 180°E longitude meets 180°W (with just a few jogs so that

78

island groups aren't divided). Somewhere on earth this had to happen, and people agreed years ago that the best place for the days to change officially is in the middle of the world's widest ocean.

Don't worry about the lost day. If you turned the ship around and sailed east across the line again, you would get it right back. You gain a day when you cross the line heading east.

How is this possible? Time and place are related. If it were the same time at every place on earth at once, record keeping would be a lot easier. Noon in China would also be noon in Chile. But it would be bright daylight in one place and dark in the other. For that reason—because we all like our clocks to match the sun, no matter where on earth we are—the world has been divided into time zones.

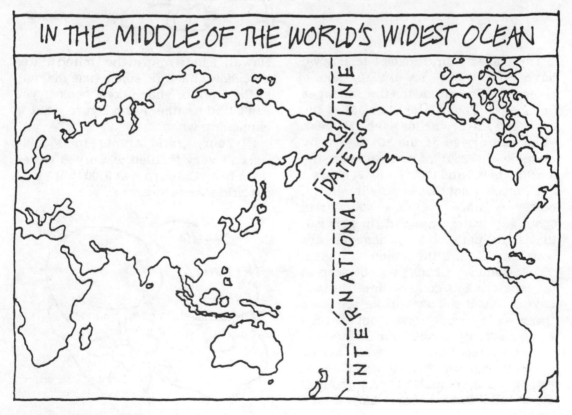

IN THE MIDDLE OF THE WORLD'S WIDEST OCEAN

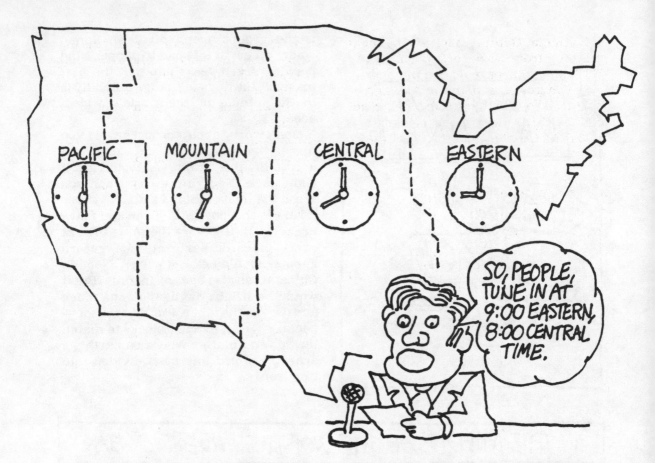

There are 24 time zones because we have divided our day into 24 hours. Every 15° of longitude the time changes by 1 hour—1 hour later each 15° to the east, and 1 hour earlier each 15° west. (You can arrive at the 15° figure by dividing the 360° of longitude in the world—180°E and 180°W—by 24 hours.)

Things are not that simple in reality, however, since political boundaries often get in the way. But, in general, that's the pattern. So when you are watching TV and the announcer says, "The Atomic Artichoke will be seen at 9:00 eastern, 8:00 central time," he isn't saying that the show will be seen two times—it's on only once, but in time zones that have a one-hour difference.

The United States spreads across seven time zones, four of which (eastern, central, mountain, and pacific) cover all states except Alaska and Hawaii. Find a map of the United States that shows the time zones (use the map in the front of your phone book if you can't find another) and figure what is happening when:

1. Your grandparents in Florida weren't very thrilled when you called them from California at 9:00 P.M., PST (pacific standard time).

2. Uncle George phoned the next morning from New York at 8:00 A.M., EST, and your parents complained.

3. You got on a plane in Chicago at 3:00 P.M. by the clocks there and got off at Seattle, where the clocks read 5:15 P.M. How long did the flight really take?

4. The clock in your car (set to the local time in Denver when you left there in the morning) reads an hour earlier than those you see where you are now. Have you been traveling west or east?

Answers: 1. Your grandparents were probably in bed—9:00 P.M. in California is midnight in most parts of Florida, and 11:00 P.M. in the rest. 2. It may have been 8:00 A.M. in New York, but Uncle George blew it when he called so early. It was only 5:00 A.M. in California. 3. The flight took four hours and 15 minutes total since you gained two hours traveling west. When you landed in Seattle, it was 7:15 P.M. in Chicago. 4. You must have been traveling east from the mountain time zone to the central. Clocks are an hour later in zones to the east, an hour earlier in those to the west.

CHAPTER SEVEN
Superislands

What would you see if you could look at the world as an astronaut does—from the edge of space? Mostly you would see water.

Earth is a water planet, with more than 70 percent of its surface covered by oceans and seas. It seems almost a mistake to call our world "Earth," since so little of it is actually solid ground.

Some of the land areas are tiny islands, specks that you could hardly see from way up there, hundreds of miles above the earth. Others are huge chunks of land that stretch for thousands of miles: the superislands that we call continents.

Fits and Pieces

"I think I've found something over here," shouts Judy to the others in the crew. She carefully brushes the dirt away from the smooth, shiny surface of a piece of pottery she has found buried in the Arizona desert. The others continue the search, and by the end of the day, more than 20 pieces have been found nearby.

These bits and pieces of clay are brought back to the laboratory, where they are painstakingly fitted together. Even though a few small chunks are missing, the archaeologists—scientists who specialize in digging up the past—can see that they have found a beautiful bowl made by the Anasazi Indian peoples more than 800 years ago.

The young German scientist studied his maps thoroughly. He felt that the fit between the two continents seemed too good to be a mere accident. If he cut South America from the left side of the map and turned it just a little, it fit almost perfectly into the coast of Africa.

To Alfred Wegener it looked as if the continents had once been joined together and had somehow drifted apart across the Atlantic Ocean to the places where they are now. Maybe, just maybe, he thought, they were once joined like a bowl that has since been broken apart. In his mind he put the bits and pieces back together, and it seemed to him that they fit very well.

Scientists have long agreed that it is all right to reconstruct pottery from fragments found in the same vicinity. But until recently they have rejected the idea that the continents might have once fit together. Most scientists felt certain that Wegener had been working too many jigsaw puzzles. Surely

continents of solid rock could not have broken apart and wandered thousands of miles from where they had formerly been joined together. Preposterous!

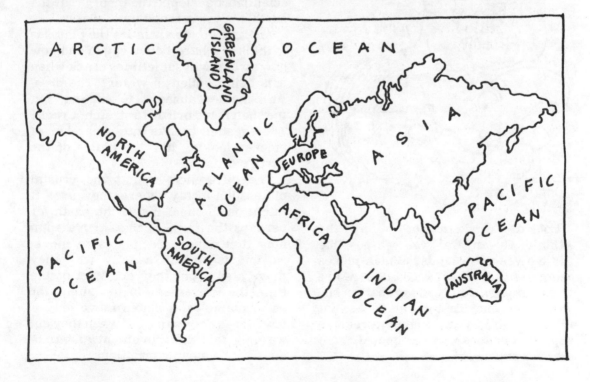

Today, however, most earth scientists agree that Wegener's farfetched idea that the continents were once joined together is essentially correct. Many clues besides the good fit between the pieces led them to believe that the giant superislands have moved around over the face of our planet.

Look at a world map or a map of the Atlantic Ocean. Can you see possible fits between the lands? Which pieces seem to fit together neatly? Where are there gaps? You can probably see where Wegener first got his ideas, but it was many years before proof to support his ideas was collected from all over the world.

Plates That Go Bump in the Night

The idea of drifting continents began to make much more sense when scientists developed the theory that the earth's crust is made up of plates.

An easy way to visualize the plates is to think of a hard-boiled egg. You know how an egg will sometimes crack when it is being boiled in water? The pressure inside causes it to expand, but the shell is too brittle to stretch. Crack! The eggshell breaks into a number of pieces—looking like the plates of the earth's crust.

The difference is that the crustal plates don't stay where they are. It might be the heat inside the earth, or perhaps the force of the earth's spinning that causes the plates to move. No one is certain exactly why they move about, but they do. Most of the time the movement is too slight for us to notice—the plates move only a few inches per year—but when they do, we may feel the movement as earthquakes or volcanic eruptions.

The real action takes place at the edges of the plates. In some places—such as in the middle of the Atlantic Ocean—the plates are spreading apart. Plates in other areas are being squeezed together.

In fact, some plates are pushed or pulled so hard that they are forced underneath other ones and are finally melted deep inside the earth.

Some of the big chunks of crust carry only ocean on their backs, while some of them carry the huge blocks of light rock that we call continents. Most plates carry some of each.

Where the plates carrying continents bump into each other, tall mountain ranges are often formed—the crust is squeezed together, and mountains are shoved up in towering ranges. In other places, plates carrying continents collide with oceanic plates. The result is that the oceanic plate is forced under the continent, creating deep trenches and arcs of islands.

If all this sounds too difficult to

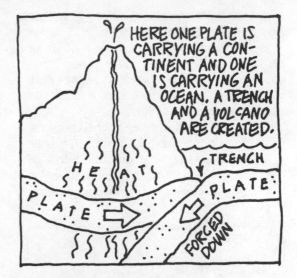

HERE ONE PLATE IS CARRYING A CONTINENT AND ONE IS CARRYING AN OCEAN. A TRENCH AND A VOLCANO ARE CREATED.

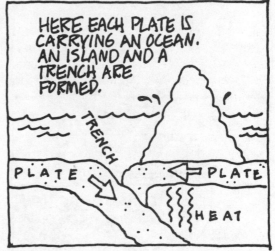

HERE EACH PLATE IS CARRYING AN OCEAN. AN ISLAND AND A TRENCH ARE FORMED.

imagine, you can see how it works for yourself by making your own plates. You'll need some stiff paper—index cards, or even paper plates, should work—and a blob of clay (play dough or dry, stiff mud) to serve as your continents.

First try the plates (paper) without continents (clay) on them. If you are using paper plates, turn them upside down. As you slide the plates together, they should either slip past each other—one going under the other—or collide. What if they collide? If you push them together slowly, you can

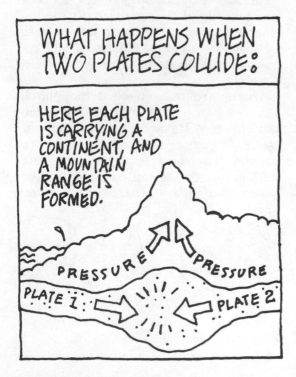

WHAT HAPPENS WHEN TWO PLATES COLLIDE?

HERE EACH PLATE IS CARRYING A CONTINENT, AND A MOUNTAIN RANGE IS FORMED.

feel tension increasing as they buckle upward or downward.

What happens if you keep pushing one against the other? The same thing that takes place when two oceanic plates meet head on—one eventually is forced under the other.

Let's see what happens when a plate carrying a continent meets another without one. Put a good-size hunk of clay (or whatever you are using for your continents) on one plate and bring them together again.

HEY, MOM. SEE HOW GEOGRAPHY WORKS?

If they don't collide, and the continental plate overrides the oceanic one, they slide on without much trouble. (In reality the continental plate also has a blob of rock just as deep on the bottom as on the top. The plate diving underneath scrapes this rock, with earthquakes and volcanic activity resulting.)

What if the continental plate starts to go under the oceanic plate? Try it and see. You may be able to push your plates hard enough to move your continental blob, or you may find that your plates act the same way the earth's crust does. The continental rock forces the oceanic plate to flip-flop and turn under the plate carrying it.

See what happens when two plates carrying continental material come together. The plates carrying them are bent and squeezed together until one is forced down. Then the blobs run into each other, and there occurs a dramatic folding of the crust—something your soft clay probably won't do. The earth's highest mountains are believed to be the result of crashes between continental blocks carried on the backs of plates that have crunched together.

Where are the edges of the plates that go bump in the night—and the daytime too? If you trace the centers of the earthquakes and active volcanoes around the world, you will find the outlines of the dozen or more plates that make up the crust of our ever-changing planet.

LOOKS LIKE WE'RE IN FOR ANOTHER CHANGE.

RUMBLE

North America's Icy Past

Take a look at a map of the North American continent and notice that there are more lakes in the northern half than in the southern half. A lot more.

Why do you think this is? Those thousands of northern lakes are clues to the past, proof that North America used to look very different from the way it does now.

As strange as this may sound, the story of the continent's past can be seen in the freezer compartment of your refrigerator. Take a look inside at how the frost builds up to the point where there is hardly room for frozen foods and ice trays. The warm, moist air of your kitchen is touching the very cold surface of the freezer compartment and crystallizing as ice (unless you have a frost-free refrigerator, which moves the air too quickly for ice to form).

Most of the northern part of North America was covered with ice sheets in the past. No one knows for certain why, but about a million years ago the earth's climate turned cooler. Snows that fell in the winter didn't melt during the summer, but began piling up in deep drifts. Eventually those compacted until the snow was crystallized into icy glaciers—rivers of ice—in the far north and high mountains.

You can see how snow is turned to ice by squeezing a snowball in your hands. The pressure changes your soft snowball into a ball of ice.

Thousands of years passed, and the glaciers grew together into giant ice sheets. These spread southward from their starting points in central Canada until more than 60 percent of the entire continent was covered by ice at least 100 feet thick.

Four times the great ice sheets covered the land, and four times they retreated. As recently as only 10,000 years ago, the ice sheets began to puddle for the last time, in the areas that later became the valleys of the Columbia, Missouri, Mississippi, Ohio, and Hudson rivers. Some isolated glaciers had even blanketed the southern Rocky Mountains as far south as New Mexico, and the Sierra Nevada to within 200 miles of present-day Los Angeles.

Glaciers and great ice sheets leave footprints, which knowing geodetectives find. As the glaciers pushed their way slowly southward (moving perhaps as much as one football field, or 300 feet, per year), the heavy ice gouged the earth. It picked up boulders weighing many tons just as easily as it moved countless small pebbles.

Water from the melting of the ice sheets supplied the land it had already carved with lakes. The present Great Lakes, large as they are today, are only remnants of the former fresh-water seas formed during the melting period. If you look again at the map of North America and see Lake Winnipeg in the Canadian province of Manitoba, you will find a small portion of what was once Lake Agassiz. It covered practically all of southern Manitoba, extended into eastern Saskatchewan and western Ontario, and even flooded south into Minnesota and North Dakota.

The ice invasion left traces of its passing throughout the northern United States in the form of rocks picked up along the way. Some of the big boulders found on Long Island, New York, originally sat on the ground in Canada hundreds of miles away.

In the following paragraphs are descriptions of additional evidence found in North America that shows the work of those icy fingers of an earlier time.

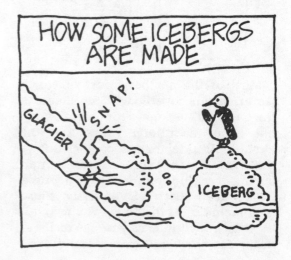

There are still some glaciers in the mountains of Alaska and Canada today. The rivers of ice in the St. Elias Mountains reach down to the sea; and when they break off, they form icebergs. Can you find Glacier Bay, which is named for its icy surroundings?

The climate of North America was much cooler during the Ice Age, and there were large lakes in the western United States, where none or only smaller ones exist today. The largest of these was Lake Bonneville, which once covered an area that is mostly salt flats now. The Great Salt Lake is a small remnant of North America's icy past.

The Arctic Ocean coastline of Canada and Alaska was not covered by the ice sheets and glaciers. This strange occurrence came about not because it wasn't cold enough there, but because there wasn't enough moisture for heavy snowfall.

Middle America, Where the Earth Shakes

Imagine yourself in a Mexican cornfield on a sunny afternoon in February 1943. As you watch farmer Dionisio Pulido working, you notice that a thin stream of smoke is rising from the ground. The earth begins to rumble and fine ash begins to accompany the

smoke. There is a smell in the air that reminds you of rotten eggs.

By daylight the next morning the small hole in the ground has changed into a cone of smoking cinders 30 feet high. The earth continues to rumble and shake as smoke, ash, and large rocks are thrown into the air. This noisy outpouring continues all day long, building an even larger cone of cinders that reaches 150 feet above the surrounding fields.

The events of the next day are even more startling. Thick chunks of molten rock come pouring out of cracks near the base of the growing cone, covering

the cornfield to a depth of more than ten feet. The hissing of steam and the roar of the explosions from inside the cone are deafening to anyone standing within a mile, and smoke billows upward thousands of feet into the darkening sky.

Within a few weeks the molten rock flows and ash have buried not only the cornfield, but Dionisio's home village of Paricutin (pa-ree-ku-TEEN).

That little puff of smoke you saw was the birth of a volcano 1,500 feet high, whose lava flows buried more than 15 square miles of farmland.

The area between the continents of North and South America has seen many sights similar to this one. That long, narrow strip of land known as Central America is the place where several of the earth's crustal plates are pushing together. It was there in the centuries before the arrival of Columbus that the Aztec, Maya, and Toltec peoples built their great cities of stone.

The records these native civilizations left tell us that their lands have been shaking for thousands of years. A line of volcanoes runs the length of Central America, from just north of Paricutin, south through Panama, including the towering twin peaks of Popocatepetl (po-po-caw-tay-PET-il), which is 17,887 feet high, Citlaltepetl (seat-lol-tay-PET-il), which is 18,701 feet high, and fiery Izalco (ee-ZAHL-co) in El Salvador.

THE HOME OF A BUNCH OF ACTIVE VOLCANOES

Volcanoes are no strangers to the islands of the West Indies either. A few of the islands in the group called the Lesser Antilles have active volcanoes that have proven destructive in this century, the most famous of which is Mt. Pelee (PAY-lay) on the island of Martinique (mar-ti-NEEK).

Pelee had been sending rumbling signals for several months before the big eruption took place, and worried islanders from all over Martinique had come to the town of St. Pierre to seek refuge. But on the morning of May 8, 1902, the side of the volcano blew open, and a cloud of superheated steam and ash roared down toward the town. Within minutes the fiery cloud engulfed St. Pierre, killing the entire population of 30,000 people, except one scorched prisoner who was in a cell below ground.

The same crustal plate movements that produce volcanic activity have also produced many devastating earthquakes in Central America. During the 1970s alone, two earthquakes destroyed the capital cities of Managua, Nicaragua, and Guatemala City, Guatemala. These earthquakes left more than a million and a half people homeless.

The "land bridge" connecting the two Americas is thus a shaky one. Yet some of the land's restlessness has been helpful to people because the soils produced by volcanic activity are unusually fertile. Many of the bananas and coffee beans enjoyed by people around the world would not grow so well had it not been for the scary events that gave the region its rich soil.

What about the area where you live—has there been a history of volcanic activity or earthquakes? We ordinarily think of volcanoes and earth-

quakes occurring only on the western edge of North America, where the earth is very restless. The Cascade Range in California, Oregon, Washington, and British Columbia has shown that it is far from extinct—a term best reserved for retired volcanoes. Mt. St. Helens in the Cascade Range erupted explosively in 1980.

Most people who have studied the geologic, or earth science, history of the West expect strong earthquakes there again at any time.

Even if you don't live in the West, you may be able to find some shaky

A RETIRED VOLCANO

earth in your area's background. Surprisingly, powerful earthquakes have been known to shake such places as Boston, Massachusetts; Charleston, South Carolina; and the Mississippi Valley. See if you can find out if there has ever been a quake in your area, and if so, when it happened.

There may even have been volcanoes in your area's past. The granite rock that is found in many parts of the eastern United States was formed by volcanic action within the earth's crust many years ago, and there are other places where ancient lava once poured out of the earth to form features that can be identified even today. Do you have any remnants of a fiery eruption in your own backyard? What is that smell? Whew, it's only some eggs boiling.

(the midline of the planet, which gets more direct sun than any other place) and find yourself in a snowfield and yet have hot steam rising all around you?

Sounds impossible? It would be, unless you happened to be on top of steaming, volcanic Cotopaxi in Ecuador, South America. Although this peak is close to the equator, the nearly 20,000-foot elevation of Cotopaxi has a snowy cap just like its nonvolcanic neighbors.

This Ecuadorian volcano is just one of hundreds of tall peaks that make up the longest continental mountain range in the world, the Andes. The South American chain—also called a *cordillera,* the Spanish word for "range" —stretches the entire length of the continent, from its northern edge, on the Caribbean Sea, to the southern tip of Cape Horn. That's more than 5,000 miles!

As you look at a map of South America, you can see several striking things about the rugged backbone of that continent. Notice that the Andes are very close to the Pacific (west) coast of South America. Standing on the top of Cotopaxi, you would be less than 150 miles from the ocean, and 3½ miles high.

AN EARTHQUAKE-RESISTANT HOME

The Amazing Andes

Here's a riddle for you. How could you be only 30 miles from the equator

Perhaps the most dramatic point from which to see just how steep and high the Andes are is a place in northern Chile. Llullaillaco (you-ya-e-YA-ko) is an ancient volcano that is over 22,000 feet above sea level, and yet just 80 miles off the coast is an ocean trench that is more than 26,000 feet deep. The difference in elevation between these two points is thus greater than 48,000 feet—nine miles—even though they are only 200 miles apart.

Scientists now think there is a good explanation for the steepness of the Andes, and their location so close to the seacoast. The plate carrying South America is moving westward, over-riding the Nazca plate, which is headed eastward. The resulting pressure of the plate being forced downward under the continent has been strong enough to create a deep trench offshore and to thrust the Andes high above the rest of South America.

To us, mountain ranges such as the Andes seem enormously high, but they are merely tiny bumps on the surface when compared to the size of the whole earth.

Balloon Skins

Here's an experiment for you to try that will give you some idea of how the tallest mountains compare with the size of the earth. Find a round balloon and blow it up until it has a diameter of about ten inches. (Diameter is the distance across a round object, measured through the center.)

If your balloon now represents the size of the earth, the tallest mountains on its surface would not rise even the thickness of the balloon skin above sea level.

Another way to compare mountains to the size of the earth is to pretend the earth's diameter is one mile and to walk "through" the center of the earth.

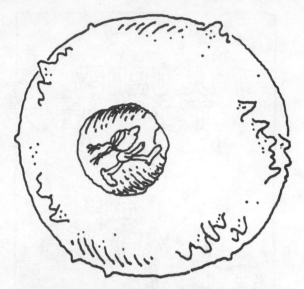

If the diameter of the earth were one mile, you would only have to walk three feet more to represent the height of 22,831-foot-tall Mt. Aconcagua (ah-con-COG-wah), the highest peak in the Andes.

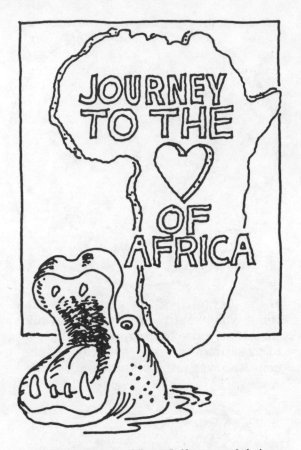

channels, instead of a single mouth. Ancient Greeks saw that the shape of the land between the outermost channels looked like the triangular shape of a letter in their alphabet and called it the "delta." Today this term is used not only for the Nile, but to describe the areas at the mouths of rivers where stream-carried silt forms rich plains.

Africa is considered the most interesting of the world's superislands for good reasons. It is the second largest continent and a naturalist's paradise, with a greater variety of plant and animal life than any other continent.

Mystery is a part of Africa, too, because its interior isn't an easy place to know. Great natural barriers prevented outsiders from taking a look at the heart of the continent until the last hundred years.

Africa has great rivers that penetrate into the interior. Three of the world's longest rivers are in Africa. African rivers can be traveled by boat for only a few hundred miles, so its secrets were long protected from the curious eyes of strangers.

The longest river on earth is the mighty Nile, which empties into the Mediterranean Sea through a series of

More than 4,000 miles from where it empties into the sea, the Nile begins its life in the snow-capped, volcanic Mountains of the Moon and the high Mufumbiros. There in the heart of Africa are found the tallest and the shortest peoples on earth—the Watusi and the Mountain Pygmies, respectively—bamboo and banana forests, baboons and chimpanzees, lions and buffalo.

If Africa seems very distant, you'll probably be surprised to know that you may have a bit of the Nile in your pocket. Take a look at the back of a one-dollar bill. Notice the circle on the left-hand side of the bill. Just below the shining eye is something from the land of the Nile. What is it doing there and what does it mean? That's another African mystery.

THE CLIMATE GETS COLDER THE CLOSER YOU GET TO THE POLES

The Not-So-Cold Continent

Everyone knows that the climate gets colder the closer you get to the poles. In the Northern Hemisphere, that means things get colder the farther north you go. The continent of North America generally illustrates this quite well, especially along its east coast. Can you rank these cities according to how cold they are in winter, with the coldest first and the warmest last?

Philadelphia
Jacksonville, Florida
Norfolk, Virginia
Boston
Baltimore
Portland, Maine
Miami
New York City
Columbia, South Carolina
Washington, D.C.

If it gets colder as we get closer to the pole, the continent we call Europe should be the coldest of those inhabited

---

Answers: (with average January temperatures) Portland (22.4° F.), Boston (28.9), New York City (32.3), Philadelphia (33.1), Baltimore (33.2), Washington (36.1), Norfolk (41.6), Columbia (46.6), Jacksonville (55), Miami (67.5).

by humans. Its geographic center is closer to the North Pole than any other. But when we compare the average winter temperatures of European cities with those of North America, on the opposite side of the Atlantic Ocean, we get a big surprise.

Look at Washington, D.C., the capital of the United States, and Paris, the capital of France. Paris is about 700 miles closer to the pole than Washington, D.C., is, so it should be a colder place, right? Well, it's not. An average January day in Paris (37° F.) is actually a little warmer than in Washington, D.C., although the summer days are not as hot in the French capital.

We see another deviation from our general rule when we compare two cities that are about the same distance from the pole. Take Boston, Massachusetts, and Rome, Italy. Both cities are on the ocean, equally far north. But an average January day in Rome is almost 20° warmer than that average day in Boston, and summer days are hotter too.

How can this be? It seems that the oceans are responsible for the variations.

Did you know that there are rivers in the ocean? That's right, there are currents that flow along the surface of the seas very much like rivers flow on land. Some of these are cold ones, which cool and dry the coastlines they pass by. And some, like the warm North Atlantic Current, bring warmth and moisture to the land.

From what oceanographers now know, it appears that the warmth that reaches the west coast of Europe starts off the hot west coast of Africa. The sun warms the water, which is blown westward by the trade winds. This warm current travels all the way across the Atlantic without so much as a suitcase, picking up warmth from the equatorial sun. So it is called the North Equatorial Current.

When it runs into the northeast coast of South America, the current is partially funneled into the Caribbean Sea, through the Gulf of Mexico, and finally back into the Atlantic through the Straits of Florida. (Are you following its journey on your map?) Then it is

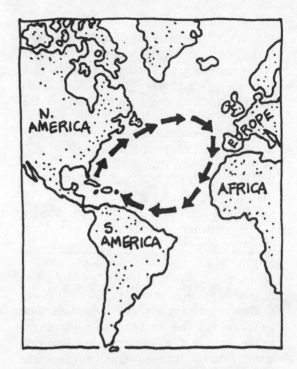

called the Gulf Stream, and it links up with other currents and heads northeast up the coast of North America. Although it changes its name to the North Atlantic Current along the way,

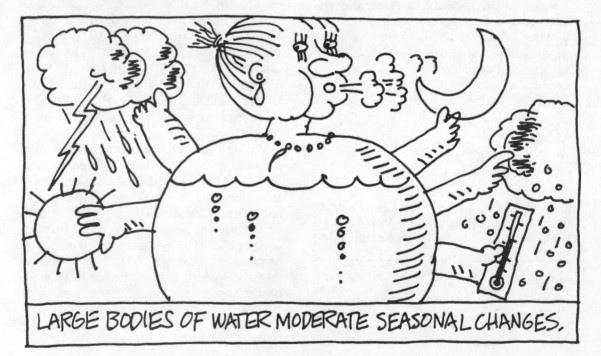

LARGE BODIES OF WATER MODERATE SEASONAL CHANGES.

this warm surface stream keeps flowing until it reaches the west coast of Europe. Presto. Europe gets a welcome blanket of warmth, keeping it far warmer than the other side of the Atlantic.

There is something else at work too. Looking at a map of Europe, you can't help noticing that it has a very irregular coastline—seas and bays cut into it everywhere. As a result, in most parts of Europe there is a strong marine influence on the weather. That means there aren't the extremes of temperature that "continental" areas of the earth have, since large bodies of water keep the lands they touch from getting very hot or very cold. You can see that almost no place in Europe is more than a few hundred miles from some sea or large bay. (Some geographers don't consider Europe a continent at all. They consider it just a big peninsula at the western end of the continent they call Eurasia.)

This closeness to the sea has had a great influence on the peoples of Europe as well as on their weather. Many of them learned to use the seas as the road to profitable trade and even to piracy. The seafaring Vikings of the Scandinavian lands of northwestern Europe were perhaps the boldest of them, sailing far and wide 1,000 years ago. These adventurers conquered the British Isles and portions of France, Spain, and Italy. They even sailed west to settle in Iceland and Greenland, and were probably the first Europeans to set foot in North America.

These seagoing spoilers of earlier days wouldn't have even been able to chip their long boats out of the cold northern ice if it weren't for the warm current. It gave their lands warmer winters than Minnesota—1,000 miles farther south—where modern Vikings trudge over frozen football fields.

Since the ancestors of many Americans came from Europe, that continent has contributed a lot to the American language. For example, think about the names of dogs. You will come across places in Europe where those names came from.

1. A big dog, the Great _____, was named after the country that is located between the North and Baltic seas.

2. A well-known dog with long, red hair is the _____ Setter, named after one of the British Isles.

3. High in the snowy Alps you might find the huge _____ with a flask of whiskey around its neck. This breed was named after a famous Alpine pass.

4. This white dog with a lot of black spots was named _____ after the Adriatic sea coast of central Yugoslavia.

5. The _____ is well known as a military and police dog from a country in north central Europe.

6. These large white dogs were named _____ after the mountain range that separates the Iberian Peninsula from the rest of Europe.

7. A small dog with shaggy black hair and bushy eyebrows has been named a _____ because its home was in the northern part of Great Britain.

8. The hot dogs some American kids love so much are sausages called _____ after a German city.

9. Speaking of sausages, can you find the names of the city in northern Italy, the capital of a central European country, and a northern European country that have sausages named after them?

GUESS WHERE THE NAME FOR HAMBURGER CAME FROM.

Answers: 1. The Great Dane was named after Denmark, the peninsula and island country. 2. Those red dogs are Irish Setters, named for the island and nation of Ireland. 3. St. Bernards are the big dogs that are known for their rescues in the snowy Alps near the St. Bernard Pass, which connects Switzerland with Italy. 4. Yugoslavia's Dalmatian Coast gave its name to those spotted dogs who supposedly like to hang around firehouses. 5. With its pointed ears and nose, the German shepherd is better known than its shaggy-coated cousin, the Belgian shepherd. 6. Pyrenees (sometimes called Great Pyrenees) are large, strong dogs sometimes used to pull carts in the mountainous region that separates Spain from France. 7. Scotties, or Scottish Terriers, are named after Scotland, which is north of England. 8. Hot dogs are also known as frankfurters, a name they got from Frankfurt, Germany. (Bet you can't guess where the name for hamburger came from.) 9. Sausages? Well, there's that great Italian bologna (from Bologna, Italy, of course), not to mention Vienna sausages from Austria's capital city (called Wien there), and the spicy Polish sausage from—where else?—Poland. Is all this talk of food making you Hungary? Norway! Too much Greece.

Most of All — Asia

Asia is the continent with the most of practically everything. It is the largest continent—containing nearly a third of our planet's land surface—and has most of the people—two-thirds of the human population live in Asia. The highest and lowest points on earth are there, and it has places that are about as wet or dry, as hot or cold, as any other continent.

As large as it is, Asia is naturally a continent of many contrasts. So although it has lands that are the most densely populated on earth, it also has vast areas where few, if any, people live. The island of Java in southeast Asia is about the same size as the state of New York; yet it has more than four times as many people. There are areas of the Arabian Peninsula in southwest Asia that are equal to twice the size of New York State; however, they have a human population of zero.

The thing that is most remarkable geographically about the giant continent is its rugged heartland. Unlike the other continents, the heart of Asia is so mountainous and dry that it has kept its peoples from having much contact with each other.

All you have to do is look at a map of Asia to see the great barriers. High mountains run in an almost continuous band from the Aegean Sea in Turkey through Iran, then on through the higher Hindu Kush and Pamir ranges in Afghanistan and the towering Karakoram Mountains of Kashmir. Then come the mighty Himalayas, the "roof of the world," topped by 29,028-foot Mt. Everest, which appears to be still growing!

North and east of the Himalayas are the high plateaus of Tibet and Mongolia, some of the toughest territory on earth to travel. Any communication and trade between areas around the outside of this core required crossing snowy passes, barren deserts, and wild mountain rivers on narrow trails.

This isolation led the many peoples of Asia to develop their societies in different ways, and those differences give the continent a colorful kaleido-

OVERLOOKING THE GREAT BARRIERS OF ASIA

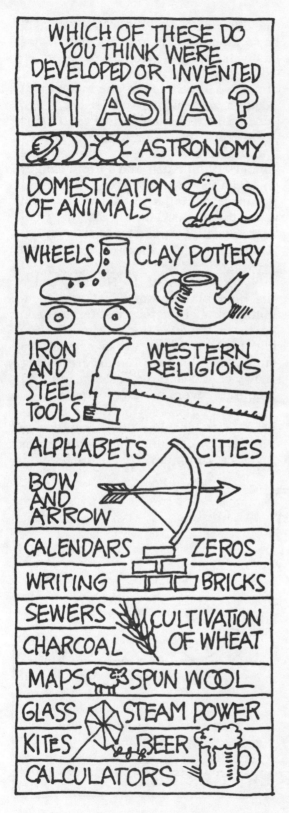

WHICH OF THESE DO YOU THINK WERE DEVELOPED OR INVENTED **IN ASIA?**

ASTRONOMY

DOMESTICATION OF ANIMALS

WHEELS   CLAY POTTERY

IRON AND STEEL TOOLS   WESTERN RELIGIONS

ALPHABETS   CITIES

BOW AND ARROW

CALENDARS   ZEROS

WRITING   BRICKS

SEWERS   CULTIVATION OF WHEAT

CHARCOAL

MAPS   SPUN WOOL

GLASS   STEAM POWER

KITES   BEER

CALCULATORS

scope of cultures even today.

Asian peoples have enriched our Western societies more than most of us know. Some of the things that are a part of our daily lives can be traced back to origins in Asia.

Which of the following do you think were developed or invented in Asia? Astronomy; domestication of animals; wheels; clay pottery; iron and steel tools; Western religions (Christianity, Islam, Judaism); alphabets; cities; bow and arrow; calendars; writing; zeros; sewers; bricks; charcoal; cultivation of wheat; maps; spun wool; glass; steam power; kites; calculators; beer.

If you guessed all of them, you are right. They all originated in Asia—a continent of giant inventions.

The paper this book was printed on, for instance. Did you know that paper was invented in China nearly 2,000 years ago? And about a thousand years later a man named Pi Sheng experimented with baked clay and invented the first movable type. The printing press used in Europe didn't come along for nearly another 500 years.

Down Under

Australia is the world's smallest continent—and the most unusual. For one thing, it is "down under," located entirely within the southern half of the globe. You know, where people have to stand on their heads and everything is upside down. At least, that's what a lot of people used to think things were like there.

Even though we now know that people stand on their feet everywhere in the world, Australia does have very odd plants and animals when they are compared with those found on other continents.

What else but odd could you call a six-foot-long, striped "dog" with a tail that never wags, a "fox" that flies, a five-foot-tall bird that weighs over 100 pounds and lays dark green eggs (and doesn't fly), or a fur-covered creature

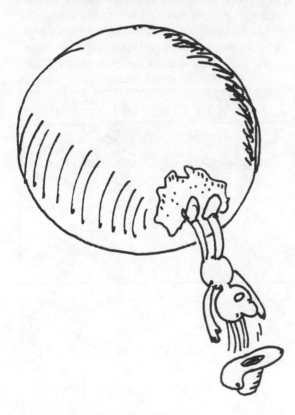

with a tail like a beaver's and a bill like a duck's that lays eggs and gives milk to its young?

Have you ever heard of any of these animals? Do you know their names? They are the Tasmanian wolf, the giant fruit bat, the emu, and—strangest of all—the duckbill platypus.

How did things get to be so different there from other parts of the world? Two things seem to have combined to make Australia and its neighboring islands different from other continents—isolation and a lack of water.

Australia is remarkably flat as continents go. There are some upland areas, but only along the eastern fringes do they rise high enough to get seasonal snowfall. This lack of dramatic elevation combined with generally dry-wind conditions gives most of the continent a very dry climate. The coastal areas get moderate rainfall, but most of the interior is either desert or near-desert.

More than a million square miles—a third of the whole continent—averages less than ten inches of rainfall per year. Ten inches sounds like a lot of rain, since that much would fill your bathtub just about to the overflow

RAIN?

rain at all, while occasional storms may drop several year's average moisture in a single downpour that floods miles of land for several weeks. The rain that does fall won't be around long—Australia is close to the equator and the sun shines very hot.

Isolation is the other thing that seems to have made life in Australia different from other continents. It is not only the smallest of the continents, it is separated by hundreds of miles of ocean from its nearest neighboring continent (Asia). All the other continents are connected by land, except at the Bering Strait. This narrow gap of shallow water separating Asia and North America has been bridged many times in the past when sea levels have dropped.

So the plants and animals of Australia have developed on their own for the past 40 to 60 million years, with only a few intrusions from outside.

One of the intruders was man. No one knows exactly how long ago it was when the ancestors of the Australian

drain. But when you figure that's all you would get over the entire year, it's not much.

Averages don't give the whole picture of just how dry Australia's "outback" really is. Some years there is no

DEVELOPING ON THEIR OWN FOR 60 MILLION YEARS

aborigines (a-bo-RIDGE-uh-knees) arrived on the continent, but it was probably during the last great glacial period. The lowering of ocean levels and the push of other peoples from Asia may have led the aborigines into Australia from the big island of New Guinea at some time between 20 and 50 thousand years ago.

What they found was a land that looked pretty much the same as it does today, with somewhat more rainfall. These people adapted their society to the increasingly drier conditions of the continent so well that when European explorers first came 200 years ago, they found aborigines living in practically every corner of the land.

These people had little in the way of material possessions to show the curious explorers. But they had learned to live on the scanty native vegetation; to hunt the kangaroo, wallaby, and emu; to avoid the poisonous snake and wolf; and most of all, to find water.

Hunting kangaroos is not an easy thing. The largest of these powerful leapers is the red kangaroo, which weighs up to 200 pounds and is very fast. How fast?

Well, how fast can you run 10 yards (30 feet)? Give yourself a running start and see if you can cover the 30 feet in two seconds. If it takes two seconds, you are running at just over 10 miles per hour. The red kangaroo can cover the same distance in only two-thirds of a second — 30 miles per hour.

Kangaroos can move fast because of the big jumps they take. A red "boomer" going full speed can just about cover the 30 feet in one jump! How far can you jump with a running start? If you can jump half as far as a kangaroo, you are not a bad jumper.

The Deep-Frozen South

Do you ever have days so bad that you just wish you could be left completely alone? Well, there is one continent where you could get your wish — nearly six million square miles without another human living there. But you would have to put up with a few inconveniences.

The nearest store would be quite a walk — 600 miles or more. And the temperatures get a little cool. Several readings of less than -125° F. have been

guin. This flightless sea bird stays through the winter, rearing its young in places where the temperature rarely climbs above -50° F. The penguins look surprisingly formal in their feathery tuxedos of black and white against a pure white backdrop of snow and ice.

Antarctica doesn't keep all that cold to itself. Quite a bit is shared with the oceans in the form of icebergs. These are gigantic chunks of ice that break off from the continental sheet and slip into the ocean. They drift hundreds or even thousands of miles before they finally melt, cooling the seas in the southernmost parts of the world.

You can create something like this effect at home using ice cubes. To make it more realistic, add a teaspoon of salt to one glass of water, and fill another glass with plain tap water of the same temperature. Then put a cube of ice about the same size into each glass.

Notice that the cubes float, just like the crushed ice you get in a soda. This is very curious because most of the time hot substances tend to rise and cold ones to sink. You may even be able to see this happening if you look closely

recorded, but summer days sometimes warm up to around zero. There aren't any trees around for firewood (since the soil is frozen all year around and the largest plants that can grow are mosses), but there is coal in places.

The problem is that most of the coal and everything else is buried beneath a sheet of glacial ice that is more than a mile deep—more than two miles thick in some spots.

About the only land visible on the continent of Antarctica is the tops of the highest mountains and a few isolated miles of partially thawed shoreline. While the sun never sinks below the horizon in December, it never comes up at all in June and may only shine for a few hours in May and July.

As tough as this sounds, there are some hardy animals that make the icy continent their home. Seals and a few other sea mammals climb out of the cold waters surrounding Antarctica at times to sun themselves on the ice pack.

Probably the best-known inhabitant of the continent is the emperor pen-

at the bottom of the cubes and watch the small stream of particles dropping down toward the bottom of the glass. You may also be able to see the warmer water near the outside rising toward the top.

Why do icebergs (and your artificial ones) float on the surface? They float because of the strange properties of water. Unlike most other substances, water expands as it freezes. That means that it weighs less than the nonfrozen water around it, and it floats. If not, the floors of the oceans would contain only cold icy sheets instead of the many forms of life they do support.

Notice that ice doesn't weigh a lot less than the water—icebergs are mostly under water. Like your cubes, only a small portion of their total size shows on the surface. Can you tell which one is floating higher, the tap-water or the salt-water cube? Salt water is a little heavier than fresh water, so those icebergs in the sea get a little extra boost.

If the earth's temperature were to rise slightly, as it has done in the past, the icecap covering Antarctica might melt completely. So much water is locked up in the ice there that if it all melted, sea levels around the entire world would be raised by at least 100 feet.

105

history. Greenland, the largest island, seems to be just that—a stopping-off place for sailors of the stormy North Atlantic. It has fewer than 50,000 people living in its vast, cold expanse, but it plays an important role in transportation even in the jet age.

Even more could be said for the four island nations of Indonesia, Japan, the United Kingdom, and the Philippines. Almost a tenth of the world's people live there, although they add up to only one-fiftieth—2 percent—of the land area of the earth. Largely because of their locations, they have developed into trading centers for the world.

How would this affect your town? Would any of your state be under water? (Don't lose too much sleep over this question—it would take thousands of years for all the ice to melt, we are told.)

Did you notice anything about the way your cubes melted? Did one melt faster than the other? What happens if you add new cubes to each glass?

Stepping Stones

In addition to the continents, which account for most of the world's land areas, there are thousands of smaller islands. Some of these are not so small —the 6 largest add up to more than two million square miles (over half the size of the United States). The 46 largest islands are all bigger than the state of New Jersey.

There are some islands that sit off by themselves in the middle of oceans, but most of them are located in groups (called "chains" or "archipelagos") and are close to the continents.

That slight distance from the mainlands has given these islands important positions as the stepping stones of

1. South of Florida, this long, narrow island is the home of ten million Spanish-speaking people.

2. These islands off the coast of Morocco share their name with a famous, yellow, singing bird.

3. The name of this North Atlantic island would better suit its frozen neighbor, Greenland.

4. Italy seems to be giving this big Mediterranean Sea island the boot.

5. The world's fourth largest island is

separated from Africa by the Mozambique Channel and has a vehicle and fuel in its name.

6. Some of the finest teas come from this big island that sits only a few miles off the coast of India.

7. Indonesia and Malaysia share parts of the world's second largest island.

8. Japan's largest island is the home of its capital city and highest mountain.

9. This "down under" country, 1,200 miles southeast of Australia, is made up of two large islands.

10. The island state of the United States is made up of 132 large and small volcanic islands in the North Pacific.

Answers: 1. Cuba, a Spanish colony until 1898, is a close neighbor of the Florida peninsula. 2. Columbus stopped off in the Canary Islands on his way to explore the New World. 3. Although called Iceland, Greenland's smaller neighbor has steaming geysers and active volcanoes. 4. Sicily is the island that the Italian peninsula appears to be kicking around. 5. This is the island where you would be *mad* if you ran out of *gas* in your *car*—Madagascar. 6. Sri Lanka (also called Ceylon) is a tea-producing island nation of 15 million people. 7. Borneo is divided between Indonesia, Malaysia, and the tiny country of Brunei. 8. Volcanic Mt. Fuji and the city of Tokyo are both located on the Japanese island of Honshu. 9. New Zealand is made up of two large islands of nearly equal size. One is called North Island; the other, South Island. 10. Hawaii is the name of the state and largest island in the chain.

CHAPTER EIGHT
The Big Puddles

To most of us the oceans are just big puddles. We think of them as the barriers that separate the really important areas of the earth—the continents and islands we live on—from one another. If it weren't for the fact that we use the seas to carry our goods from place to place and that we depend on them for part of our diet, most of us could ignore them.

But the salty waters that cover most of the surface of our home planet are more important and interesting than that. Beneath those waves lies the story of our planet and the life on it. We really can't say that we know much about this place where we live without first knowing something of those great puddles.

So let's leave our solid-earth home and take a plunge into the seas that hold many of the mysteries of life in our world.

Probably the first person who drifted to sea on a log many thousands of years ago wondered what lay beneath the waves. Since then, nearly everyone who has seen the ocean has asked: How deep is it, and what can be found there?

Answers to these questions have been a long time coming. Early peoples soon learned many things about the surface of the seas—they understood wind and waves, how to navigate by the sun and stars when out of sight of land, and how to take food from the oceans. But we haven't learned very much about the undersea world until the past few decades.

Most ancient peoples who thought about the oceans at all figured that they were very much like teacups. Their sides sloped gradually away from the shorelines toward nearly featureless plains at the bottom. Sailors from the earliest periods of recorded history noted that the shallow waters near ports gradually gave way to the dark colors of the deeper seas, and these observations were backed up by the use of weighted anchor chains and fishing lines, which were used as measuring tools.

Sounding a Puddle

You can use this old but very reliable method yourself. Have you ever found a puddle that looked deep, but you couldn't see the bottom because it was too muddy? You might have attacked the problem head-on by wading in to find out. Instead you could have sounded the depths with a string line. To do so you need a good length of string, a heavy rock, and a tape measure or yardstick—and, of course, some watery depths to measure.

You can try out something larger than a puddle—a lake, a creek, or a pond will probably be more interesting. The important thing is to find a place from which you can get yourself over the spot to be sounded. A low bridge or dock will work well since either can get you right over a deep spot.

IT'S IMPORTANT TO FIND A WAY TO GET OVER THE SPOT TO BE SOUNDED.

Tie your rock (or some other heavy object that doesn't float) to the end of your line. You can make your sounding line a little fancier by measuring convenient lengths (every foot, for example) on the line above your weight and marking them with knots or waterproof ink. That way, you'll have some idea of how deep your weight has dropped, even while you are lowering it into the water.

Okay, now you are ready to take a reading. Lower the line slowly, feeding it out a little at a time. If there is a strong current, you may have to use a

109

heavier weight—otherwise your line may be pulled away from where you are standing as it sinks. You should be able to feel when the weight reaches the bottom because the line will become loose in your hands. (The string may also begin to float.)

Now you can pull up the line. Look carefully for the point at which it first becomes wet. By measuring the distance between this and the weight, you will have found the depth. Some string absorbs water quickly, so the moisture will continue to climb up the string. That's why you should find the beginning of the wet part as quickly as possible.

Try other places nearby to see if they are about the same depth. If you mark the first depth on your string with a knot or ink, you will be able to tell if a nearby spot is shallower, or deeper,

and estimate how much just by looking. If other places are deeper, the string will be wet higher up. Are you getting a picture of the bottom you can't see?

A note of caution: Be sure that you do this in a safe place and that you are careful as you take your readings. After all, measuring with your body's length isn't the idea. Neither is measuring from a bridge so high that you need a hundred feet of string just to reach the water.

And if the game warden reminds you that it's not fishing season yet, be sure that you don't have a hook on the other side of the weight. You're conducting an experiment, remember?

BE SURE YOU TAKE A READING IN A SAFE PLACE.

Deeper Than Deep

Deep-sea measurements of the Atlantic Ocean were not at first made to satisfy scientific curiosity, but to survey a route for the first underwater telegraph cable in the 1850s.

The results were surprising. Many of the soundings revealed that the Atlantic Ocean floor was more than 15,000 feet deep in many places, but they also showed undersea mountain peaks reaching to within a few thousand feet of the surface. The seabed turned out not to be as smooth as a teacup, but every bit as varied as the landscapes of the continents.

In this century, researchers have found that the great ocean basins of the world are generally between 12,000 and 18,000 feet deep (about two to three and a half miles). But they have also discovered a curious thing: In some areas there are trenches that are much deeper. How deep? Well, the deepest trench discovered so far is 37,000 feet (seven miles) deep. Now, that's a lot of string!

The kid is fearless. Even though there could be deadly water snakes, hungry crocodiles, and toe-eating sharks waiting for him, he steps forward into the shallow end of the municipal swimming pool. Last night on TV he saw some guy wrestle with a giant octopus, and he saw another save a whole town from being eaten by a lobster. If the heroes of "Demons of the Deep" weren't scared of the water, he isn't going to be either.

The water at the bottom of the steps is only two and a half feet deep – sissy stuff, thinks the kid. He keeps going past the smaller kids playing in the shallow part of the pool and reaches the tile that is marked "3 Ft."

Still he pushes forward. The bottom of the pool is sloping gently toward the dark blue waters of the deep end, and the kid passes the tile marked "3½ Ft." Now it's time for him to get ready for the big plunge. He adjusts his Frogman Freddie flippers and pulls his Aquaman swim mask over his face.

By the time he gets to the "4-Ft." mark, the kid is up to his neck in water. He tiptoes ahead until he can barely hold his head above the wet stuff, reaching for the rope at the "5-Ft." mark. This is it, he thinks. His toes can feel that just underneath the rope the gradual shallow slope changes to a much steeper one.

He takes a deep breath, lets go of the rope, and sinks toward the bottom. Through his foggy mask he can see the terrifying depths of the deep end, where the big kids send huge waves all over the pool with their cannonballs and Egbert almost empties the pool with his bellyflops. The water is dark down there at 16 feet, and that slope down is pretty scary too. The kid comes up gasping for air.

"Wutzamatter, kid, see a shark?" says the lifeguard. "Maybe you'd better stick to the shallow end until you learn how to swim."

So what does the Shallow-End Kid have to do with the oceans? Just that the same thing happens there as happens in your ordinary swimming pool. The continents are surrounded by gently sloping continental shelves, the shallow ends that slope gradually to the much deeper ocean basins.

90 PERCENT OF THE WORLD'S SEAFOOD COMES FROM CONTINENTAL SHELVES.

Of course, the shelves are a lot deeper than 5 feet. They usually end somewhere between 400 and 800 feet deep, and they vary quite a bit in size. The continental shelf extends several hundred miles out from some places, such as from the southeastern coast of South America, but only a few miles out from the continent's steep southwestern coast.

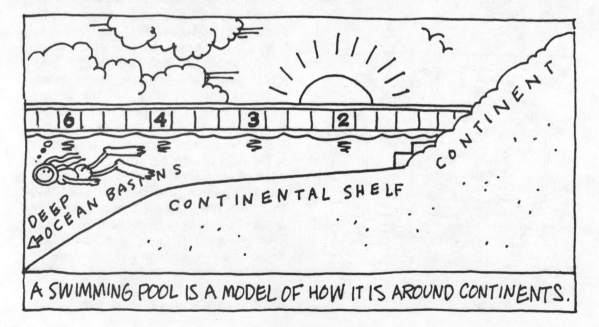

A SWIMMING POOL IS A MODEL OF HOW IT IS AROUND CONTINENTS.

These shallow waters are very important to all of us because of the tasty resources located in them. More than 90 percent of all the world's seafood comes from the continental shelves and shallow seas located close to land. These areas are abundant with fish life, including important bottom-living species: cod, halibut, flounder, and haddock. Fishing is our oldest use of the seas, and the worldwide shortage of protein foods makes it even more important today.

Here are descriptions of some of the world's leading fishing grounds. See if you can find and name them.

1. Shallow sea between Great Britain, Holland, Denmark, and Norway.

2. Two seas off the east coast of China, north of the island of Taiwan, south and west of Korea.

3. Banks off the southeast coast of Newfoundland.

People are continuing to find other resources beneath the shallow waters. Oil and gas are frequently found in the coastal areas of the earth, and drilling rigs now dot the shallow waters of most continents. And there are plans to harvest metals and other mineral wealth from continental-shelf regions. What this means is that there will be a lot of work for the Shallow End Kid — when he learns to swim.

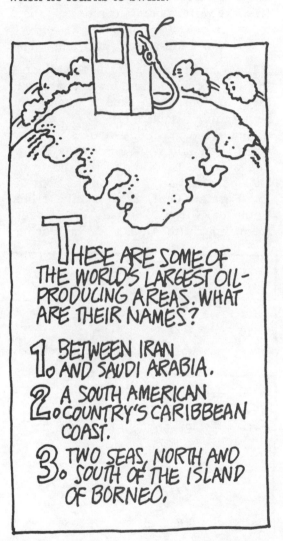

Answers: 1. North Sea — also a major oil and natural gas field of Europe, 2. Yellow Sea and East China Sea, 3. Grand Banks of Newfoundland.

Answers: 1. Persian Gulf — world's largest known reserves of oil, 2. Venezuela — the Gulf of Venezuela and Lake Maracaibo areas, 3. South China Sea and Java Sea — both part of the Sunda Shelf.

Even though the continental-shelf areas are relatively shallow, they are among the most dangerous places in the world to work. Storms whip waters into monster waves in the shallow North Sea throughout the year, and strong currents plague the nearby English Channel. Anybody who has ever been near the Gulf of Mexico when a hurricane came through can tell you just how mean those usually friendly waters can become.

Pulling the Plug

If we could somehow pull the plug on the world's oceans and lower the water level 600 feet, there would be another seven million square miles of new land added to the continents and as islands. Then you could walk across the North Sea to Great Britain and Ireland from Europe, stroll anywhere on the floor of the Persian Gulf, take a hike from China or Korea to Japan, cross over from Siberia to Alaska on the Bering Sea land bridge, or ride a bus from Boston 300 miles east to the seashore.

You would also be able to find a lot of junk! For thousands of years people have been dumping things in coastal waters. Parts of the Mediterranean would be littered with ancient wine bottles, and Long Island Sound would be filled in places with rusted cans. But there would also be some treasures—many ships have sunk in the shallow waters of the world, including those grand Spanish galleons loaded with gold.

A Note on Names

Have you ever found yourself wondering why some bodies of water are called oceans and others are called seas, some bays, and others gulfs, sounds, or channels?

Salt-water names are somewhat like land names—they don't always have reasons for being called one thing and not another. So a narrow body of water

BODIES OF SEA WATER

STRAIT

PASSAGE

CHANNEL

NARROW

GULF

BAY

SOUND

PARTIALLY ENCLOSED

OCEAN

SEA

BIGGEST

There are four bodies of water called "oceans." A fifth, the Southern Ocean, is recognized by some geographers. It is made up of the southern portions of the other oceans and has no well-defined boundaries.

Oceans are set apart from other bodies of water by their size. The Pacific, Atlantic, Indian, and Arctic oceans together account for 93 percent of the earth's surface waters. But no matter what a body of water is called, as long as it isn't totally enclosed by land, it is part of the great world ocean.

WET BASEMENTS

If you think your basement is cold, damp, and dark, imagine what it must be like at the bottom of the ocean. No light penetrates to the cold depths of those basins, where the temperatures stay around 30° F. all year round. (They don't freeze because salt water doesn't freeze until it reaches 28° F. or colder.)

And the pressure at 30,000 feet is incredible—13,500 pounds (nearly seven tons!) per square inch. At sea level the pressure is only 14.7 pounds per square inch. So the forces pushing on every square inch at the inky depths of

can be called a "strait," a "channel," a "passage," or such-and-such "narrows." A partially enclosed body of sea water may be a "bay," while another one nearby may be called a "sound."

The only real exception to this random pattern seems to be the oceans.

30,000 feet under the ocean would be the same as having an entire football team of 54 men, each weighing 250 pounds, piled on top of you.

For those reasons most of what we know about the floor of the deep seas comes to us secondhand. Instruments such as echo sounders, magnetometers, and thermocouples have told us most of what we know about the ocean depths. But there have also been deep-sea explorations carried out by people in submersible tanks called "bathyscaphes," which have carried them to the bottom of the deepest trenches. What would a trip to the ocean floor be like?

UNDERSEA TRIPS IN A BATHYSCAPHE

Imagine that you are leaving San Francisco for a voyage in a submarine to the deepest part of the ocean. The harbor is only 30 feet deep at the pier from which you leave, and there is a lot of oozy mud on the bottom. That changes a lot by the time you reach the narrow channel known as the Golden Gate. There the swift tidal currents that flow between San Francisco Bay and the Pacific Ocean have scoured the floor to a depth of more than 300 feet.

Outside the bay the floor of the sea is covered with a layer of silt (soil and rock particles) washed downstream by northern California rivers. Visibility there is improving because the silt is settling on the bottom, but the increasing depth of the water cuts out more and more sunlight.

This is the continental shelf, which extends more than 60 miles from the coast. To your left (the south) are the Farallones Islands near the edge of the shelf. They are the last points of land which rise above the surface for the next 2,000 miles of your journey.

Just beyond the Farallones lies the continental slope, which falls off sharply from a depth of around 700 feet to the real floor of the Pacific at around 18,000 feet beneath the surface. The slope is rugged and uneven as it takes you ever deeper. At the 8,000-foot mark you can see (with the aid of strong lights) a huge chunk of rock called the Pioneer Seamount rising toward the surface, not quite making it — it is still 2,400 feet from the surface when it ends.

Beneath you now is a yawning submarine canyon with jagged features. It is called the Murray Seascarp and marks a deep fracture in the earth's crust. The depth here is more than three miles (16,000 feet). Turn off the strong lights for a moment so you can see one of the strangest sights of the ocean — glowing fish. Some of the sea creatures that live in these depths have become adapted to the inky darkness by providing their own glow.

With the lights back on you now see the true bottom of the Pacific for the first time. You have traveled more than a thousand miles from San Francisco. Ahead there is more rugged sea bottom, not the smooth flat plain that people expected to find only a few years ago. Surprisingly, even at this

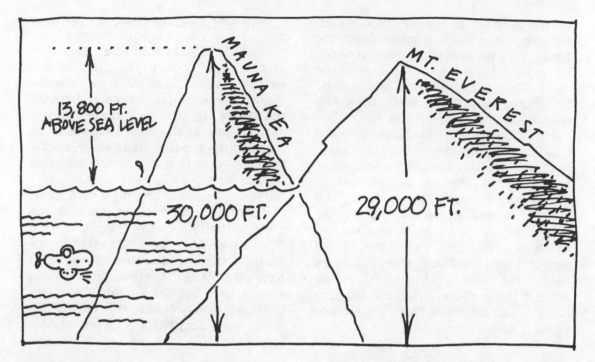

depth, many animals are living at or near the bottom.

A bit later your undersea ship has to head back toward the surface because you are approaching the volcanic Hawaiian Islands. You pass through a channel northwest of the major islands about 15,000 feet down. The islands you have bypassed include the world's tallest mountain, Mauna Kea.

World's tallest mountain? But Mauna Kea is only 13,796 feet above sea level—not nearly as high as 29,028-foot Mt. Everest in the Himalayas. True, but Mauna Kea rises from an ocean basin that is more than 16,000 feet deep. The summit of its volcanic dome is therefore more than 30,000 feet above its base. The piles of lava it took to produce the great cone of Hawaii add up to 10,000 cubic miles—more than enough to cover the entire state of New York with 1,000 feet of the stuff.

For the next 3,500 miles your submarine travels through a series of canyons, over ridges, and past seamounts. You have crossed several spots in the basins 20,000 feet deep, but they hardly compare to the gaping hole you see before you now.

This is the Mariana trench. It is only part of a 4,000-mile-long gash in the earth's crust that extends all the way to the Bering Sea. But here at the

western end of the trench is a hole so deep that you could fit Mauna Kea in it and still be more than a mile from the surface of the ocean.

Even at the bottom of this ocean trench—the deepest spot now known on earth—your strong lights reveal life. Flatfish, shrimp, and strange luminescent animals in the viewing ports show you that animals live even in the bottom of the very dampest, darkest "basement."

Heading Back Home

Hey, you've come a long way from your own house. All the way to the bottom of the ocean, in fact. But now at least you know how to find your way back home.